Combined Arms Warfare in Israeli Military History

Dedication

This book is dedicated to my uncle,

Paul Nathanson.

Our many conversations about the IDF over the years will always be a source of fond memories for me.

Combined Arms Warfare in Israeli Military History

From the War of Independence to Operation Protective Edge

David Rodman

SUSSEX
ACADEMIC
PRESS
Brighton • Chicago • Toronto

2 4 6 8 10 9 7 5 3

First published in hardcover 2019, reprinted in paperback 2020, in Great Britain by
SUSSEX ACADEMIC PRESS
PO Box 139
Eastbourne BN24 9BP

SUSSEX ACADEMIC PRESS
Independent Publishers Group
814 N. Franklin Street, Chicago, IL 60610

British Library Cataloguing in Publication Data
A CIP catalogue record for this book is available from the British Library.

Library of Congress Cataloging-in-Publication Data
Names: Rodman, David, author.
Title: Combined arms warfare in Israeli military history : from the War of Independence to Operation Protective Edge / David Rodman.
Description: Brighton ; Chicago ; Toronto : Sussex Academic Press, [2019] | Includes bibliographical references and index.
Identifiers: LCCN 2018046022 | ISBN 9781845199678 (hbk : alk. paper) | ISBN 9781845199807 (pbk : alk. paper)
Subjects: LCSH: Israel—History, Military | Israel. Tseva haganah le-Yisrael—History | Unified operations (Military science).
Classification: LCC DS119.2 .R63 2019 | DDC 355.4/6095694—dc23
LC record available at https://lccn.loc.gov/2018046022

MIX
Paper | Supporting responsible forestry
FSC® C013604

Typeset & designed by Sussex Academic Press, Brighton & Eastbourne.
Printed and bound by CPI Group (UK) Ltd, Croydon, CR0 4YY

Contents

List of Abbreviations

AAA	Antiaircraft Artillery
AEW	Airborne Electronic Warfare
ALA	Arab Liberation Army
A'MAN	Military Intelligence Branch (Israel)
APC	Armored Personnel Carrier
ARM	Antiradiation Missile
ATC	Air Transport Command (Israel)
AW	Asymmetrical War
AWACS	Airborne Warning and Control System
C^2	Command and Control
C^3I	Command, Control, Communications, and Intelligence
CAS	Close Air Support
CAW	Combined Arms Warfare
CBU	Cluster Bomb Unit
ECM	Electronic Countermeasures
FPB	Fast Patrol Boat
GHQ	General Headquarters
GP	General Purpose (Bomb)
HQ	Headquarters
IADS	Integrated Air Defense System
IAF	Israel Air Force
IDF	Israel Defense Forces
IED(S)	Improvised Explosive Device(s)
IN	Israel Navy
IW	Interstate War
LIC	Low-intensity Conflict
MOSSAD	Foreign Intelligence Service (Israel)
MTB	Motor Torpedo Boat
NATO	North Atlantic Treaty Organization
NSA	National Security Agency (United States)
ORBAT	Order of Battle
PA	Palestinian Authority
PFLP	Popular Front for the Liberation of Palestine

PGM	Precision-guided Munition
PLO	Palestine Liberation Organization
RDB	Runway Destruction Bomb
RMA	Revolution in Military Affairs
SAM	Surface-to-air Missile
SHIN BET	Domestic Intelligence Service (Israel)
SNS	Special Night Squads
SOF	Special Operations Force(s)
SPEC OP(S)	Special Operation(s)
SSM	Surface-to-surface Missile
TRADOC	Training and Doctrine Command (United States)
UAV	Unmanned Aerial Vehicle
UN	United Nations
USAF	United States Air Force
USN	United States Navy
WMD	Weapons of Mass Destruction

List of Illustrations

Preface

Combined arms warfare (CAW)—the integration of different arms on the battlefield in an effort to achieve maximal efficiency there—is as old as war itself. Every army across both time and space that has engaged in combat has practiced one version or another of CAW, whether consciously or otherwise. In the preindustrial era, CAW involved the integration of, inter alia, chariotry, cavalry, heavy and light infantry, and artillery. In the industrial era, it has involved the integration of, inter alia, armor, mechanized and nonmechanized infantry, artillery, aircraft, and engineers.

The Israel Defense Forces (IDF) has certainly been no exception to the rule in respect of the practice of CAW. Throughout six interstate wars (IWs), three asymmetrical wars (AWs), numerous low-intensity conflicts (LICs), and untold special operations (SPEC OPS), the IDF has engaged in CAW in one form or another. Nevertheless, though the IDF is by now the subject of an enormous literature, no single work expressly traces its experience with CAW from its establishment in 1948 to the present day. This book, therefore, is intended to be a first attempt to fill a significant lacuna in the literature about the IDF and Israeli military history.

Specifically, this work examines each of the IDF's six IWs, each of its three AWs, each of its LICs, as well as a sample of its SPEC OPS, explicitly from the perspective of CAW. Furthermore, the book not only evaluates the IDF's triumphs and setbacks on the battlefield, but also traces the evolution in the IDF's approach to CAW throughout the past seven decades.

Before dipping into the topic, however, I would be remiss if I did not express my gratitude to certain individuals and organizations for their direct and indirect contributions to this work. First, I would like to thank Aharon Klieman and Stuart Cohen, both of whom sharpened my own thinking about the IDF in the context of an entirely different project about Israeli national security. Any problems in respect of either fact or interpretation in this book, of course, are solely my responsibility.

Second, I would like to express my gratitude to the journals division of the Taylor & Francis Group (<www.tandfonline.com>) for permission to reproduce here an earlier article of mine, "Combined Arms Warfare: The Israeli Experience in the 1973 Yom Kippur War," *Defence Studies* 15/2 (June 2015). Chapter 4 constitutes a revised and expanded version of this article.

And, finally, I would like to thank the Israel Government Press Office (IGPO), as well as the individual photographers, for permission to reprint photos from the Israel National Photo Collection in this book.

David Rodman
Dix Hills, New York

Introduction
The Essence of Combined Arms Warfare

"In blood and fire, Judah fell," exclaimed a renowned Hebrew poet, "In blood and fire, Judah shall rise again." These lines, which appear in Ya'akov Cahan's poem "Biryonim" ("Hooligans"), a poem composed in the early twentieth century in response to the notorious Kishinev pogrom in Russia, decades before the rebirth of the Jewish state, presciently capture not only the circumstances under which the State of Israel entered the world, but also the circumstances under which it has lived ever since its birth in 1948. Not only did the country declare its independence in the midst of a war, but also it has been engaged almost without interruption in one kind or another of armed conflict throughout its existence.

Indeed, Israel has fought nine wars to date, which averages out to a war approximately every eight years. Six have been interstate wars (IWs).[1] The other three have been asymmetrical wars (AWs).[2] Between these wars, the country has been embroiled in a seemingly endless string of low-intensity conflicts (LICs) involving on different occasions cross-border infiltration and/or border skirmishing, terrorism at home and abroad, large-scale insurgency, mass civil unrest, or some combination of these phenomena.[3] And Israel has also routinely carried out special operations (SPEC OPS) over the course of its existence.[4]

The Israel Defense Forces (IDF) has employed both offensive and defensive warfare, as well as both maneuver and attrition warfare, to protect Israel from these national security challenges, adjusting its preferred mode of combat as necessary to fit the battlefield conditions at hand.[5] During IWs, offensive maneuver warfare has been the favored mode of battle at the tactical and operational levels.[6] During AWs, offensive attrition warfare has been the favored mode of battle at the tactical and operational levels, albeit with defensive warfare occupying a more prominent niche than it has during IWs. During LICs, a more balanced mix of offensive and defensive warfare, as well as of maneuver and attrition warfare, has been the favored mode of battle. And, lastly, during SPEC OPS, the IDF has clearly emphasized offensive maneuver warfare.

One constant has nevertheless been in evidence throughout the IDF's battlefield history. Irrespective of whether it has been engaged in offensive or defensive warfare, irrespective of whether it has been engaged in maneuver or attrition warfare, and irrespective of whether it has been engaged at the tactical or operational level, the IDF has waged combined arms warfare (CAW) in one form or another. Yet, its historical experience with this dimension of warfare has thus far neither been described nor analyzed in an explicit and systematic manner. Consequently, the present volume is dedicated to an exploration of the IDF's experience with CAW across its battlefield history. First, however, it is necessary to define and explain briefly the concept of CAW, as well as to preview the contents and organization of the book.

The Concept of Combined Arms Warfare

CAW is based on the premise that an army is much more likely to be effective on the battlefield if it relies on a mix of arms (e.g., a combination of armor, infantry, and artillery) rather than on just one arm (e.g., armor alone). The United States Army defines CAW as:

> [T]he synchronized and simultaneous application of arms to achieve an effect greater than if each arm was used separately or sequentially.[7]

The coordinated use of different arms, to put it another way, acts as a force multiplier on the battlefield—and it does so regardless of whether an army favors offense or defense, maneuver or attrition.

CAW rests on two basic components: force structure and war-fighting doctrine. Force structure constitutes the "hardware" behind CAW: the armored, infantry, artillery, engineering, air, and naval formations that do the actual fighting, as well as the intelligence and logistics formations that support them. War-fighting doctrine constitutes the "software" behind CAW: the principles on which these different arms are organized and trained to work together in order to achieve maximum efficiency on the battlefield. Command and control (C^2)—that is, the management of arms on the battlefield in order to achieve a specified objective—serves as the "glue" that binds hardware and software. Talented and knowledgeable commanders—from army chief of staff down to squad sergeant—are necessary to ensure that the hardware functions in the way intended by the software. CAW, in other words, depends upon the close integration of force structure and war-fighting doctrine.

The dominant model of CAW hinges on a balanced (or symmetrical) mix of all arms: armor, infantry, artillery, engineers, and so forth. And not for no reason has this model become the traditional one employed by armies in the age of mechanized warfare, as it has consistently brought about good results on the battlefield. It is not, however, the only model of CAW that has been tested on the modern battlefield. Nontraditional models of CAW have involved the coordination of as few as two arms. The IDF's decision in the aftermath of the 1967 Six-Day War to rely on an unbalanced (or asymmetrical) mix of arms that emphasized tanks and aircraft at the expense of infantry, artillery, and engineers is a classic example of such a nontraditional model, and one examined in some depth later on.

Plan of the Book

This book consists of eight chapters. The first five describe and analyze the IDF's experience with CAW during Israel's IWs. Chapter 1 covers the 1947–49 War of Independence and the 1956 Sinai Campaign; chapter 2, the Six-Day War; chapter 3, the 1969–70 War of Attrition; chapter 4, the 1973 Yom Kippur War; and chapter 5, the 1982 Lebanon War. All but one of these chapters is organized more or less along the same lines, with an assessment of the prewar evolution of the IDF's approach to CAW preceding an assessment of the application of this approach during a particular war. The air and naval dimensions of CAW are separated out for individual consideration. Only chapter 3 deviates significantly from this structure, as the War of Attrition constituted a significant aberration in respect of the IDF's employment of CAW during IWs.

The last three chapters describe and analyze the IDF's experience with CAW during AWs, LICs, and SPEC OPS, respectively. Chapter 6 covers Israel's three AWs: the 2006 Second Lebanon War; the 2008–9 Gaza War (or Operation Cast Lead); and the 2014 Gaza War (or Operation Protective Edge). This chapter, like those that probe the IWs, provides an assessment of the IDF's prewar approach to CAW followed by an assessment of the application of this approach during war. Chapter 7 scrutinizes the IDF's conduct during the numerous

LICs in which Israel has been involved throughout its lifetime. And chapter 8 appraises the IDF's experience with CAW during SPEC OPS by examining four of the most significant Israeli SPEC OPS carried out to date.

Finally, the book includes a glossary of technical terms and a bibliography of all sources cited in the chapters.

A Couple of Caveats

First, this book does not pretend to offer a comprehensive review of Israeli military history. Rather, it dips into each of the country's IWs, AWs, and LICs, as well as a sample of SPEC OPS, only so far as is necessary to describe and analyze the IDF's encounter with CAW during that round of hostilities or during that operation. Those individuals who seek more in-depth treatments of the IDF's involvement in particular IWs, AWs, LICs, and/or SPEC OPS should consult the bibliography for further sources of information. And, second, while the general narratives of the various Arab–Israeli IWs, AWs, LICs, and SPEC OPS are not in dispute, specific facts and figures—such as the details surrounding this or that battle or operation, or the number of men and machines deployed and lost by each side in this or that round of hostilities—are often contentious. In cases of disagreement between (or among) sources, the author has adopted what he considers to be the most plausible facts and figures.

1

The War of Independence and the Sinai Campaign

Combined Arms Warfare in Its Infancy

In order to ensure its survival, the Jewish community of Ottoman- and, later, British-ruled Palestine found it necessary to develop fighting forces.[1] The Yishuv (Settlement), as the community called itself before the establishment of the State of Israel, stood up a defense group during the first decade of the twentieth century. Known as HaShomer (The Watchman), its task was to protect Jewish agricultural villages from bands of Arab marauders bent on pillage and destruction. In the decades after the British government's promulgation of the 1917 Balfour Declaration, which recognized the right of the Jewish people to rebuild its "national home" in its ancestral land, Arab violence against the Yishuv became steadily more organized, widespread, and bloody. In order to cope with this new reality, the Jewish community replaced HaShomer with Haganah (Defense), an organization that would evolve over the same decades into a large and efficient militia, complete with an elite force of "shock troops" known as the Palmach (an abbreviation for Strike Companies). The Haganah's policy of self-restraint in the face of Arab violence eventually spawned two dissident militias—the Irgun Zvai Le'umi (National Military Organization) and the smaller Lohamei Herut Yisrael (Fighters for the Freedom of Israel), the former commonly referred to as the Irgun and the latter as Lehi—that adopted much more aggressive, tit-for-tat approaches in their responses to Arab violence.[2]

The Haganah, Irgun, and Lehi were unsanctioned (by the British authorities), clandestine militias and, therefore, chronically short of weapons. At the end of the British Mandate for Palestine in mid-1948, their arsenals contained only light mortars, machine guns, small arms, and improvised explosive devices (IEDs)—and all of these weapons in very limited quantities. From the perspective of force structure, in other words, none of the militias possessed the capability to wage combined arms warfare (CAW), either traditional (symmetrical) or nontraditional (asymmetrical), in the manner of a fully fledged army, though the Haganah did learn to coordinate the use of mortars, machine guns, small arms, and IEDs to good effect in small infantry teams. The Haganah acquired its knowledge of CAW at the tactical level in part through trial and error during engagements with Arab insurgents and, in the wake of the Second World War, British soldiers. But most of its war-fighting doctrine grew out of its experience cooperating with the British army during the 1936–39 Arab Rebellion in Palestine, when many of its members joined local British forces in fighting Arab insurgents, and during the Second World War, when many of its members joined the British army, some of them as part of the Jewish Brigade that fought in Italy.[3] By the time that the 1947–49 War of Independence commenced on November 29, 1947, the very day that the United Nations (UN) voted to divide the western part of Palestine—the eastern portion had been turned into the State of Transjordan by the British government in 1922—into separate Arab and Jewish states, then, the Haganah at least had a substantial cadre of war veterans with some knowledge of CAW.

Combined Arms Warfare in the First Phase of the War of Independence

The War of Independence had two distinct phases: (1) an intercommunal phase that lasted until mid-May 1948, during which Jewish and Arab militias fought each other for control over western Palestine's territory during the waning months of the British mandate, and (2) an interstate phase that lasted until early 1949, during which the newly established Israel Defense Forces (IDF) fought the invading armies of the surrounding Arab states, which sought to crush the Jewish state at birth. The War of Independence, to put it differently, was partly a low-intensity conflict (LIC) and partly an interstate war (IW).

During the intercommunal phase of the war, the Jewish militias mustered perhaps 40,000–50,000 men (and women), of whom approximately half were capable of field operations, the rest being restricted to static defense of Jewish towns, villages, and neighborhoods.[4] Only a fraction of this manpower was equipped with weapons, mainly small arms. Cooperation among the Haganah, Irgun, and Lehi was loose at the best of times and often nonexistent. The Arab militias mustered similar numbers of men, again mostly equipped with small arms. Palestinian Arabs comprised the bulk of the membership of these militias, but the Arab Liberation Army (ALA), comprised primarily of Iraqi and Syrian "volunteers," played a prominent role in the hostilities during this phase of the war. Furthermore, the Arab militias could count on the surrounding Arab states for matériel support in respect of arms and ammunition, whereas the Jewish militias had no external allies, except for largely powerless Jewish communities around the world, on whom they could count for matériel support. British army forces, ostensibly responsible for security in Palestine until the end of the mandate, largely refrained from intervening in the fighting; however, when they did so, it was usually for the benefit of Arab forces, consistent with the British government's anti-Zionist policy of the day.

At the outset of the War of Independence, the Yishuv was on the defensive, the local Arab community on the offensive. On the one hand, the Haganah, already organized into brigades by this juncture, proved to be a rather efficient force in its defense of Jewish towns, villages, and neighborhoods, effectively integrating mortars, machine guns, small arms, and IEDs for this purpose; consequently, the Arab militias made scant headway in overrunning Jewish population centers, frequently suffering heavy casualties in their assaults. On the other hand, the Haganah was not nearly as efficient at keeping open the roads that linked these towns, villages, and neighborhoods. Arab militias were frequently able to cut vital supply lines between these population centers, risking not only the territorial integrity of the Yishuv, but also the survival of isolated towns, villages, and neighborhoods whose continued existence depended on maintaining a physical connection to the rest of the Jewish community.

Hence, Haganah general headquarters (GHQ) decided that its forces must go over to the offensive to secure the Yishuv. Operation Nachshon and Plan D were the fruits of this decision.[5] The objective of these operations was twofold: to relieve pressure on isolated portions of the Yishuv and to liberate as much of the territory assigned to the Jewish state under the UN partition resolution as possible before the end of the British mandate and the expected Arab invasion. Throughout April and the first half of May, Haganah forces, with some assistance from the Irgun, made significant headway in achieving these goals, in part by employing traditional CAW at the tactical level in their offensive operations against Arab strongholds. Mortars, machine guns, small arms, and IEDs were integrated during infantry assaults. Furthermore, homemade armored vehicles—civilian trucks or buses with armored plates welded to them—were employed in support of the infantry. Nevertheless, despite the fact that the Jewish militias had clearly gained the upper hand against their Arab foes by the termination of the British Mandate for Palestine, the War of Independence was far from won.

Combined Arms Warfare in the Second Phase of the War of Independence

On May 14, 1948, the last British army forces withdrew from western Palestine, terminating Great Britain's 30-year rule over the area. Israel declared its independence on the same day. The armies of five Arab states—Egypt, Jordan, Iraq, Syria, and Lebanon—immediately and simultaneously invaded the new state, heralding the interstate phase of the War of Independence.[6]

Upon the establishment of the country, the Israeli government transformed the Haganah into the IDF, which later absorbed Irgun and Lehi forces into its ranks as well. The newly formed army initially had more or less the same order of battle (ORBAT) as the Jewish militias from which it had sprung. It had 40,000–50,000 men (and women) in its ranks.[7] Approximately half of these soldiers, organized into nine brigades, were capable of field operations, while the rest were adequate solely for static defense. The IDF possessed enough light weapons—mortars, machine guns, small arms, and IEDs—to equip only about half of its manpower. It had no heavy weapons to speak of—no fighter-bombers, no tanks, no purpose-built armored or engineering vehicles, no artillery, and no warships—though such arms would begin to trickle into its arsenal soon after the country's declaration of independence, having been acquired abroad earlier by the Yishuv from both Eastern (through official channels) and Western (through unofficial channels) bloc sources. Such weapons could not, of course, be delivered to Jewish forces so long as the British retained control over Palestine.

For its part, the Arab coalition of states and militias arrayed against the IDF had a similar number of fighting men capable of field operations and static defense. Contrary to popular thought, the Arab coalition never had a significant manpower advantage during the interstate phase of the war. Rather, the coalition's great advantage, at least at the outset of the interstate phase of hostilities, lay in its possession of substantial quantities of heavy weapons, including fighter-bombers, tanks, armored vehicles, and artillery. Moreover, some of these weapons were the most modern versions available on the arms market.

Three rounds of fighting—punctuated by two temporary, internationally brokered truces—took place during the interstate phase of the war. In the first round of fighting, the Arab coalition, because of its great advantage in heavy weapons, was generally on the offensive, the IDF generally on the defensive. Though coalition forces registered some territorial gains on all three fronts—Syrian and Lebanese army forces supported by ALA and Palestinian militiamen gaining ground in the north; Jordanian and Iraqi army forces also supported by militiamen gaining ground in the center; and Egyptian army forces supported by Muslim Brotherhood militiamen gaining ground in the south—IDF ground forces were nevertheless able to contain the Arab offensive before the imposition of the first cease-fire on June 11. They did so by making defensive stands across the country—defensive stands that efficiently integrated mortars, machine guns, small arms, and IEDs, as well as the first artillery tubes and antitank weapons to arrive from abroad. A handful of Israel Air Force (IAF) fighter-bombers even made an appearance on the southern front, helping to stop the Egyptian army in its tracks.

On those occasions when IDF ground forces went over to the offensive during the first round of fighting, however, they did not fare nearly as well as they did on the defensive. Their setbacks here can be attributed in large measure to the poor or nonexistent practice of traditional CAW at the tactical level. Not only was cooperation within and among IDF brigades faulty as a result of poor or nonexistent command and control (C^2) of different forces, but the ground forces also failed to coordinate the employment of different arms—namely, nonmechanized infantry, mechanized infantry, and artillery—within and among these brigades to achieve their objectives. Nowhere was this woeful lack of traditional CAW more apparent than at Latrun.[8] During the first and second rounds of fighting, IDF ground forces made no less than four attempts to capture the Jordanian army fortifications in the area, which held the key to opening a road to Jerusalem, the Jewish sections of which were then under a grinding siege. The Jordanian army bloodily repulsed each attempt to capture these fortifications, inflicting the worst losses of the war on the IDF, as unsupported and uncoordinated frontal attacks by infantry forces quickly dissolved under intensive and accurate defensive fire from heavily bunkered positions.

The second and third rounds of fighting during the interstate phase of the war amounted to a mirror image of the first round in that the IDF was now generally on the offensive, Arab coalition forces now generally on the defensive. The second round of fighting, which lasted from July 9–18, saw IDF ground forces make significant territorial gains in the center and, to a lesser extent, in the south of the country, displaying in the process a marked improvement in respect of traditional CAW, with infantry being competently supported by tanks and artillery during offensive thrusts.

It would be in the third round of fighting, which lasted from October 15–January 7, 1949, however, that IDF ground forces really hit their stride in regard to the practice of traditional CAW. The IDF had built up its mechanized infantry forces by this point in the war, allowing fast-moving infantry columns—accompanied by tank, artillery, and air support—to pierce Egyptian lines in the south. In Operations Yoav and Horev, not only did IDF ground forces rout the Egyptian army out of the Negev, but also they penetrated into the Sinai, capturing territory there, withdrawing only after the British government threatened to invoke a British–Egyptian defense pact in order to attack Israel.[9] Despite a few hiccups in their performance during these operations, IDF ground forces again competently employed traditional CAW to achieve their goals. Meanwhile, in the north, IDF ground forces ejected the Syrian army, ALA, and other Arab forces, from Galilee, again in part through the utilization of traditional CAW at the tactical level.

The War of Independence was the costliest in Israeli history to date. The country lost approximately one percent of its total population, some 6,000 souls, three quarters of them combatants.[10] Arab losses are more difficult to gauge, but were probably about 10,000 souls, with roughly the same ratio of combatant-to-noncombatant dead. Arab coalition forces likewise suffered correspondingly higher equipment losses than their Israeli adversary. Furthermore, both the social and economic fabric of Israeli and Palestinian Arab society had been torn asunder. The Israelis would eventually recover from their losses to build a thriving country; the Palestinians would not do so.

Combined Arms Warfare: Air–Ground Cooperation

Airpower played a limited role in the Israeli war effort during the War of Independence.[11] At the outset of the war, the Yishuv's "air force," known as the Sherut Avir (Air Service), consisted of perhaps two dozen light civilian aircraft, whose main contribution to the war effort during the intercommunal phase of hostilities lay in their ability to deliver much-needed supplies of arms, ammunition, food, and medicine to isolated outposts. Upon the commencement of the interstate phase of the war, this task was taken over by the Air Transport Command (ATC), which also ferried into the country weapons that the Yishuv had acquired abroad. With dedicated transport aircraft at its disposal, the ATC was able to deliver supplies in much greater quantities than the Sherut Avir. Its arms and ammunition deliveries to and throughout Israel significantly enhanced the capability of IDF ground forces to continue to fight, especially during the difficult early days of the interstate phase of the war. The activation of this logistical lifeline constituted airpower's most important contribution to the Israeli victory in the War of Independence.

The ATF functioned as an independent entity during the war.[12] Simultaneously, the IAF came into existence, its responsibility being the "kinetic" dimension of airpower. Throughout the interstate phase of the war, it assembled a growing collection of Second World War fighter-bombers and even a few bombers, with which it eventually achieved air superiority over the air forces of its Arab coalition adversaries. Its command of the air, in turn, made life a bit easier for IDF ground forces, which no longer had to worry about Arab airpower. Through a combination of air combat victories and air base attacks, for example, the IAF drove the Egyptian air force completely from the sky during Operations Yoav and Horev. The IAF also flew close air support (CAS) and interdiction sorties in support of IDF ground forces throughout the interstate phase of the war, though these strikes were either very loosely coordinated or not coordinated at all with the ground forces and had only a minimal impact on land operations. The sort of air–land cooperation practiced by Allied and Axis forces alike as part of their CAW efforts during the Second World War was entirely absent from the Israeli experience in the War of Independence. Kinetically speaking, then, airpower's overall contribution to the Israeli war effort proved to be rather modest.

Combined Arms Warfare: The Naval Component

Neither the Yishuv nor the State of Israel had a navy to speak of during the War of Independence.[13] At the start of hostilities, the Yishuv's naval corps, the Palyam (Sea Companies), consisted of a few hundred men trained for sea duty and a few small boats. Its actions largely revolved around smuggling Jewish refugees and weapons past the British blockade of the Palestine coast, a continuation of the actions it had undertaken since the conclusion of the Second World War. Upon the birth of Israel, the Palyam became the Israel Navy (IN), inheriting a few rickety civilian ships that had previously been captured by the British navy in its effort to counter "illegal" Jewish immigration. These vessels were converted into warships by literally bolting completely unsuitable artillery pieces onto their decks, and they were then used for shore bombardment in support of IDF ground forces, but without effect. The only genuinely combat-capable force possessed by the IN during the war was a special operations force (SOF) equipped with explosive speedboats. This force sank an Egyptian warship and a troop transport, inflicting heavy casualties on its Arab adversary; however, this action had no apparent effect on the course of the war. The IN, in short, did not participate in any joint air–sea or sea–land operations with the IAF and IDF ground forces, respectively, and its actions did not have a significant impact on the Israeli victory in the War of Independence.

Combined Arms Warfare in the IDF Before the Sinai Campaign

Though Israel emerged victorious from the War of Independence, its survival remained far from assured in the aftermath of hostilities. The Arab world, after all, remained firmly committed to the country's eventual destruction. First, despite the fact that Israel had actually added substantially to its territory during the war, its postwar borders were still highly unfavorable from a military perspective. Extremely long in relation to the country's total land area and essentially flat, they offered no topographical obstacles to an invasion in the north, center, or south. Moreover, none of Israel's major cities, industrial assets, or military bases was located far from Arab territory. Indeed, the center of the country—which housed the bulk of its populace, most of its heavy industry, and many of its military facilities—was extremely narrow, less than 10 miles (16 km) wide in one particular spot. Not only did Israel possess indefensible borders after the War of Independence, then, but also it lacked strategic depth.

Second, numbers added another layer of uncertainty to Israeli national security. With its limited manpower base—the country's population at birth consisted of just 600,000–650,000 souls—and a correspondingly small economic base, the Israeli defense establishment realized that the IDF would not be able to compete for the foreseeable future in terms of the number of soldiers or weapons at its disposal in comparison to those in the hands of its Arab adversaries. Furthermore, most of the manpower that Israel could devote to the IDF would of necessity be made up of part-time reservists, so as not to undermine the country's economic advancement. Consequently, the Israeli defense establishment decided that it must emphasize quality over quantity, reasoning that the IDF's "few" must be organized and trained in such a way as to be able to overcome the Arab armies' "many."

Third, Israel's lack of allies contributed still further to its national security dilemma. Both Western and Eastern bloc countries sought the friendship of the Arab world, which controlled much of the globe's oil supply, because of the burgeoning Cold War. As part of their efforts to befriend the Arab countries, predictably enough, Western and Eastern bloc countries alike sought to distance themselves from Israel. This strategic reality ensured that the country could not be a party to any bilateral or multilateral defense pact and, therefore, could not count on foreign assistance in an Arab–Israeli confrontation. France and Great Britain would, of course, join Israel in attacking Egypt during the Sinai Campaign, but this development sprang from a unique set of circumstances that could not possibly have been anticipated by the Israeli defense establishment when it formulated the IDF's war-fighting doctrine in the early 1950s. Nor was such a scenario likely to recur again.

Collectively, then, these three variables—geography, numbers, and allies—drove the Israeli defense establishment, in the early 1950s, to adopt offensive maneuver warfare as the basis of its war-fighting doctrine, at least insofar as concerned interstate war. An absence of defensible borders and strategic depth made it absolutely imperative, reasoned the defense establishment, for the IDF to take the fight to Arab territory as rapidly as possible, because permitting an Arab invasion of Israel could well spell the destruction of the country. Not only must wars be fought on Arab soil, concluded the defense establishment, but they must also be short so as to prevent the Arab world from bringing its greater numbers of soldiers and weapons to bear on the fighting, so as to prevent the implosion of the Israeli economy, and so as to prevent foreign powers from intervening—even if only diplomatically—in hostilities on the Arab world's behalf. Offensive maneuver warfare, with its emphasis on rapid thrusts into the adversary's hinterland to cause its army's collapse, reckoned the defense establishment, offered the IDF the best prospect of winning wars quickly and cheaply.

The Israeli defense establishment was of one mind about the necessity of offensive maneuver warfare, but it was not of one mind in respect of traditional CAW.[14] A small "tank school" favored a nontraditional version of CAW in which armored spearheads would both pierce the adversary's defenses and exploit the break through(s) to race into the adversary's rear areas, with airpower supporting the advance of the tank columns. Infantry and artillery, according to this paradigm, would then follow behind the tanks, mopping up isolated pockets of resistance that had been bypassed by the hard-charging armored spearheads.

While the IDF's war-fighting doctrine would move substantially down this path in the aftermath of the Six-Day War, the tank school was overshadowed prior to the Sinai Campaign by a much larger "mobile infantry school," which advocated traditional CAW, with mechanized infantry taking the lead role within all-arms battle groups that included nonmechanized infantry, tanks, artillery, and engineers. The mobile infantry school won the day at least in part because of the serious teething problems experienced by the IDF tank corps during the early 1950s. The tank corps' inventory consisted mainly of Second World War-vintage vehicles that had been recovered from the scrap heap; consequently, its tanks were subject to frequent mechanical breakdowns. Based on their relative performance in the years before the Sinai Campaign, the Israeli defense establishment judged mechanized infantry to be more reliable, more flexible, and more mobile than tanks.

Consequently, in late October 1956, on the eve of the Sinai Campaign, most of the approximately 200 tanks that would participate in the war (out of a total of 300–350 in the Israeli inventory) were parceled out among the nine mechanized and nonmechanized infantry brigades that the IDF had assembled for the coming campaign.[15] The rest of the armor was concentrated in a single tank brigade. Overall, some 45,000 Israeli soldiers faced 30,000 Egyptian troops supported by a similar number of tanks. Much of the Egyptian army's ORBAT remained in Egypt proper in order to counter an expected French–British attack. The Egyptian army compensated for its inferiority in manpower, however, by occupying stout defensive positions throughout the Sinai Peninsula.

Combined Arms Warfare in the Sinai Campaign

Soon after Israel entered into a secret anti-Egyptian war pact with France and Great Britain, the Israeli government authorized the IDF to commence Operation Kadesh 2, an offensive whose aim was the conquest of the Sinai.[16] The Israeli offensive would involve four major engagements with the Egyptian army: at the Mitla Pass in central Sinai; at Abu Ageila/Um Katef and Rafah (Gaza) in northeastern Sinai; and at Sharm el-Sheikh, on the southern tip of the Sinai.

Operation Kadesh 2 began at 17:00 hours on October 29, with IAF transport aircraft dropping a parachute infantry battalion and its equipment near the Mitla Pass, in part in order to provide a pretext for a French–British offensive to seize the Suez Canal zone and in part to divide Egyptian forces in the Sinai.[17] While the parachute infantry battalion collected its equipment and fortified its position, an all-arms task force composed of additional parachute infantry, mechanized infantry, tanks, and artillery advanced across the Sinai, overcoming sporadic Egyptian resistance along the way. On the evening of October 30, this task force linked up with the airdropped parachute infantry battalion east of the pass. The next day—at the initiative of the parachute brigade commander and against the wishes of IDF GHQ—a task force composed of parachute infantry, mechanized infantry, tanks, and artillery launched a large-scale attack on Egyptian army forces defending the pass. In a long and grueling battle, IDF forces overcame their Egyptian adversary. Though the parachute brigade commander was later heavily criticized for waging an "unnecessary" battle, the relevant point here is that the IDF prevailed in the fighting as a result of its reliance on traditional CAW at the tactical level, which saw parachute infantry—the main Israeli force—supported by mechanized infantry, tanks, and artillery.

While parachute infantry descended on the Mitla Pass, an all-arms division composed of mechanized and nonmechanized infantry, tanks, artillery, and engineers began an assault on the extensive Egyptian army fortifications around Abu Ageila/Um Katef.[18] Task forces composed of tanks and mechanized infantry were able to outflank Egyptian positions in the area around Abu Ageila, capturing this part of the defensive complex by the night of October 31–November 1. IDF ground forces, however, did not fare as well at Um Katef. Egyptian forces beat back several frontal assaults on this part of the defensive complex over a two-day period, at least in part because IDF forces consistently violated the principles of traditional CAW, engaging in uncoordinated attacks that did not take advantage of the synergies created by using different arms in unison. IDF ground forces only "captured" Um Katef after its Egyptian defenders abandoned their positions as a part of a general retreat back to Egypt proper issued by Egyptian army GHQ on November 1 in response to the French–British offensive.

With Egyptian forces ousted from the central Sinai, the IDF now turned its attention to northeastern and southern Sinai. In the former sector, the IDF deployed an all-arms division composed of mechanized and nonmechanized infantry, tanks, artillery, and engineers.[19] Under the cover of artillery fire, engineers breached minefields in several places, allowing infantry and tank forces to penetrate Egyptian defenses around Rafah. Mixed teams of infantry and tanks then swiftly overcame these defenses. Traditional CAW at the tactical level had once again paid dividends for IDF ground forces. With the Rafah defenses removed as an impediment, the rest of Gaza fell to IDF forces in a series of small-scale actions that were completed by November 3.

In the latter sector, a mixed task force of mechanized and nonmechanized infantry, tanks, artillery, and engineers traveled along the coastline of the Sinai from the southern tip of Israel's Negev Desert to Sharm el-Sheikh.[20] On November 4–5, elements of this task force engaged and overcame Egyptian defensive positions at the southern tip of the Sinai in a series of small-scale actions, taking Sharm el-Sheikh and linking up with parachute infantry moving southwest from the Mitla Pass area shortly before a UN imposed cease-fire came into effect. Once more, IDF ground forces succeeded in their endeavor because they adhered to the principles of traditional CAW at the tactical level.

In about a week, aided to some extent by Egyptian army GHQ's decision to order a general retreat across the Suez Canal, in short, the IDF conquered the whole of the Sinai. In so doing, the IDF lost approximately 200 soldiers and 25 tanks.[21] The Egyptian army lost 1,000–3,000 troops and about 100 tanks.[22] Moreover, the Egyptians also left behind substantial piles of intact equipment, including tanks and artillery, some of which the IDF later added to its inventory.

Combined Arms Warfare: Air–Ground Cooperation

The IAF, assigned only a secondary role in Operation Kadesh 2 by IDF GHQ, nevertheless gave a good account of itself during the hostilities. First, it established air superiority over the Sinai, even before French–British air strikes beginning on November 1 knocked out most of the Egyptian air force on the ground. Hence, the IAF ensured that IDF ground forces did not come under effective air attack, with one exception at the Mitla Pass.[23] Second, the IAF cooperated directly, if only in a rather loose manner, with IDF ground forces by flying effective interdiction sorties of its own against Egyptian forces attempting either to intervene in or withdraw from the fighting, such as at the Mitla Pass and Abu Ageila/Um Katef. Though it is unclear how much physical damage the 490 IAF ground attack sorties inflicted on Egyptian army forces racing to or away from the battlefield, it is clear that these attacks significantly disrupted the Egyptian army's fighting capabilities by scattering its forces.[24] Third, aside from the aforementioned airdrops of parachute infantry and equipment, the IAF assisted IDF ground forces in at least two other ways. A day before the commencement of the Sinai Campaign, an IAF interceptor shot down a transport aircraft carrying members of the Egyptian general staff, emasculating the high command of the army.[25] And hours before the commencement of hostilities, low-flying IAF piston-engined aircraft used their airframes to take down telephone lines throughout the Sinai, thereby interfering with Egyptian army communications.[26] At the tactical level, then, the IAF fulfilled the roles one would expect of it in a campaign being waged according to the principles of traditional CAW.

Combined Arms Warfare: The Naval Component

Historically speaking, the IN has usually received far less in the way of resources than IDF ground forces or the IAF. Not surprisingly, therefore, it entered the Sinai Campaign with few and obsolete warships, greatly constraining its part in the hostilities. The IN did successfully provide some logistical support to IDF ground forces moving down the Sinai coastline toward Sharm el-Sheikh.[27] And it did participate in a joint operation with the IAF during the morning of October 31 that resulted in the capture of an Egyptian warship that had entered Israeli waters to fire on targets along the coastline.[28] Nevertheless, just as the IN did not have a significant impact on the outcome of the War of Independence, it did not have a significant impact on the outcome of the Sinai Campaign.

Summary

An assessment of how well Haganah and IDF forces employed traditional CAW during the War of Independence must make a distinction between fighting on the defensive and fighting on the offensive. On the defensive, these forces were generally able to integrate different arms in order to contain attacks launched by their Arab adversaries. Indeed, the Haganah and IDF's capability to wage traditional CAW at the tactical level on the defensive is perhaps the most important reason why the Yishuv and Israel, respectively, survived the difficult opening stages of the intercommunal and interstate phases of the War of Independence. On the offensive, in contrast, the Haganah and IDF's experience with traditional CAW was decidedly less positive. Haganah and IDF offensive operations often failed as a result of poor coordination between different arms and different units, the futile attacks at

Latrun being the most obvious example of this phenomenon. Moreover, even when the Haganah or the IDF displayed the capability to wage traditional CAW at the tactical level on the offensive, such as on the southern front during the final stage of the interstate phase of the war, its capability in this regard was quite rudimentary when viewed in a comparative perspective. At no point in the war did the Haganah or IDF's command of traditional CAW even begin to approach the sophistication achieved by the major Allied and Axis armies during the Second World War.

In the aftermath of the War of Independence, the Israeli defense establishment finally had an opportunity to build a force structure and think through a war-fighting doctrine suitable to Israel's strategic circumstances. This force structure and doctrine embraced the principles of traditional CAW, integrating them into a larger framework built around offensive maneuver warfare. Not surprisingly, then, the IDF's performance in respect of traditional CAW at the tactical level on the offensive during the Sinai Campaign was markedly superior to its performance in this regard during the War of Independence, though problems remained in evidence, as attested to by the setbacks incurred by its forces at Um Katef. It would not be until the next war—the 1967 Six-Day War—that the IDF really hit its stride in terms of waging traditional CAW at both the tactical and operational levels, matching or exceeding what Western armies had accomplished in the past in this respect.

2

The Six-Day War

Combined Arms Warfare Comes of Age

The State of Israel achieved one of the most impressive victories in the annals of military history during the 1967 Six-Day War—and it did so in a three-front war. Not only did the Israel Defense Forces (IDF) virtually annihilate the Egyptian, Jordanian, and Syrian air forces at the outset of hostilities, but also it systematically pulverized their armies in the fighting that followed this opening blow. In consequence of the thorough drubbing of its adversaries, the IDF captured the Sinai from Egypt, Judea and Samaria (including Jerusalem) from Jordan, and the Golan from Syria, thereby providing Israel with defensible borders and a measure of strategic depth for the first time since its birth 19 years earlier. Moreover, though the IDF's own losses in men and equipment were not insubstantial in absolute terms, they constituted a mere fraction of its adversaries' losses in the same categories.

The prevailing view of the Six-Day War in the years immediately following the hostilities was one in which the Israel Air Force (IAF) and IDF tank corps had swept all opposition before them from the battlefields. Just a couple of examples from among a great number suffice to make the point. The British military attaché in Israel during the war remarked that, "Never in the history of military aviation has the exercise of air power [sic] played so speedy and decisive a part in modern warfare." [1] A distinguished British armored corps officer observed that "the most remarkable armored operations in the . . . years of the tank's existence have been those of the Israelis in the Six Days [sic] and [latter part of the] Yom Kippur Wars."[2] Israel's adversaries shared this view. Echoing a widely held belief in the Arab world about the omnipotence of Israeli airpower, the Jordanian monarch attributed the IDF's victory against his army to the overwhelming might of the IAF.[3]

More importantly, the Israeli defense establishment bought into this view to a very considerable extent. Again, a couple of examples suffice to make the point. The IAF's chief planner of Operation Focus—the plan to destroy the Arab air forces on the ground at the start of the war—said that:

> People [that is, members of the Israeli defense establishment] said later that it was not a Six Day War but a Three Hour War, meaning the critical clash was over once Moked [Focus] had destroyed the Egyptian Air Force on the ground. Others called it a Seven Minute War, as the first surprise blow did the bulk of the damage.[4]

One of the United States Army's most prominent authorities on combined arms warfare (CAW) neatly summed up the Israeli defense establishment's attitude in the wake of the Six-Day War. "Consciously or otherwise," he argued, "[the IDF] came to rely largely on the tank–fighter-bomber team for its victories."[5] This fervent embrace of airpower and armor to the detriment of other arms would lead to serious repercussions in future rounds of fighting, particularly during the first few days of the 1973 Yom Kippur War.

The reason why airpower and armor were awarded such credit for the IDF's victory in the Six-Day War is simple: this explanation had much to recommend it. Israeli airpower and armor, after all, had assumed leading roles in the IDF's triumph. Lost in all of the hoopla about the achievements of airpower and armor, however, was the fact that the decisive ground engagements on each front—the Sinai, the Judean and Samarian, and the Golan—were decided through the application of traditional (symmetrical) CAW, not a nontraditional (asymmetrical) airpower–armor variant.

Combined Arms Warfare in the IDF Before the Six-Day War

The IDF emphasis on fighter-bombers and tanks to the detriment of other arms actually began long before the Six-Day War. Prior to the 1956 Sinai Campaign, mechanized infantry took center stage in the IDF's force structure and war-fighting doctrine. After the war, in consequence of the performance of the IDF during the fighting, the Israeli defense establishment revised its thinking. Specifically, fighters-bombers and tanks displaced mechanized infantry. In the Israeli estimation, they had become the new kings of the battlefield. Mechanized infantry—along with other arms, including artillery, engineers, and so on—took a back seat, their personnel looking on enviously as the IAF and IDF tank corps received the lion's share of the defense budget.

The IAF expanded dramatically in both quantity and quality from 1956–67. Its fighter-bomber inventory grew by over 50 percent, even more if armed training aircraft are thrown into the mix. Moreover, whereas Second World War-vintage piston-engined fighter-bombers constituted over 50 percent of its aircraft during the Sinai Campaign, the IAF had become an all-jet force well before the outbreak of the Six-Day War, thereby greatly increasing its striking power. New ordnance, such as a novel runway destruction bomb (RDB), added further to its attack capabilities. The IDF tank corps grew even more impressively over the same years. Not only did its stockpile of tanks triple by 1967, but also the vehicles in its possession were now much more modern. From an inventory that consisted largely of Second World War-vintage castoffs during the Sinai Campaign, the IDF tank corps equipped itself with modern British and American vehicles, as well as upgraded its older vehicles so that they could engage the latest Soviet models in Arab hands.[6]

The buildup of the IAF and IDF tank corps left little in the way of resources to invest in other arms. The mechanized infantry went into battle in the Six-Day War in the same Second World War-vintage half-tracks as it had employed in the Sinai Campaign.[7] The IDF made no effort to obtain modern armored personnel carriers (APCs) in the interwar years. While the artillery was largely self-propelled, the guns themselves were relatively few in number and were also mainly of Second World War-vintage. The infantry's small arms and antitank weapons, including those in the hands of the parachute infantry, were inferior to those in Arab hands. Engineers had outdated equipment as well. Finally, the IDF's logistical branch in both the Sinai Campaign and the Six-Day War relied primarily on civilian trucks, buses, and even automobiles to move men and machines to, around, and from the battlefield.

While the IDF's force structure had clearly begun to deviate from the requirements of traditional CAW by the eve of the Six-Day War, its war-fighting doctrine essentially remained committed to the concept. IDF doctrine stressed air–land cooperation at both the tactical and operational levels of warfare. Once the IAF had achieved air superiority over the battlefield by knocking out the adversary's airpower at the outset of hostilities, a task to be accomplished by a massive attack against the adversary's air bases, it would then be free to cooperate with the ground forces in their effort to smash the adversary's ground forces. IDF doctrine also stressed interarms cooperation within the ground forces at both the tactical and operational levels of warfare. The planned break-in battles in the Sinai—the only battles of the Six-Day War that the IDF had planned in advance of hostilities—envisaged, in large part, all-arms divisions composed of tanks, mechanized and nonmechanized (including parachute) infantry, artillery, and engineers piercing Egyptian army defenses.[8] Airpower, it is true, would be almost entirely absent from these battles, but only because it would be otherwise engaged in taking out the adversary's airpower. Though the "tank school" definitely had the upper hand within the Israeli defense establishment by 1967, it had yet to revise heavily the IDF's war-fighting doctrine.

The Israeli war plan for the next round of hostilities rested on an assumption—that the IDF would be permitted to strike the first blow. The deployment of IDF ground forces on the eve of the Six-Day War reflected this assumption. Most of the IDF's brigades on the Sinai front were concentrated in three divisions, each of which was located very close to the Egypt–Israel border in the area of northeastern Sinai.[9] Two of these divisions consisted of a more or less balanced mix of tanks, mechanized and nonmechanized infantry, artillery, and engineers. The third consisted primarily of tanks, because its assignment involved traversing very difficult terrain, and the IDF did not believe that its other armored vehicles would be able to keep pace with its tanks over this terrain. All three divisions were primed for an offensive maneuver campaign. Only a few independent battalions and brigades remained as a general reserve, to be allocated to the divisions as needed or to undertake separate assignments. One of these independent brigades—an armored brigade—would play an instrumental role in the defeat of Egyptian forces in the central and southern Sinai. No frontier defenses to speak of had been prepared by the IDF.

This deployment contrasted starkly with the one adopted by the Egyptian army on the eve of the Six-Day War. Though it, too, had formulated a plan to strike first as part of an offensive maneuver campaign, the Egyptian defense establishment ultimately decided to absorb an Israeli attack and then to counterattack once the IDF had worn itself out against the Egyptian army's defensive positions.[10] To this end, the Egyptian army adopted a Soviet-style "sword and shield" deployment, with infantry divisions bolstered by tank and artillery forces dug in behind stout forward defensive positions throughout northeastern Sinai.[11] Behind them, to the west and south, lay the Egyptian army's tank and mechanized divisions, ready to counterattack once the IDF had been sufficiently "softened up" by the Egyptian defenses.

Combined Arms Warfare on the Sinai Front

By June 4, the Israeli government concluded that diplomacy would not settle the latest Arab–Israeli crisis and that war had become inevitable.[12] Consequently, the government authorized the IDF to begin hostilities during the morning of June 5. Operation Focus commenced at approximately 07:00 hours. Operation Red Sheet, the ground forces attack on the Egyptian army, commenced shortly thereafter, at approximately 08:00 hours. The IDF fielded about 70,000 troops and 800 tanks in the Sinai; they faced an Egyptian army numbering about 100,000 soldiers and over 900 tanks.[13] Despite being outnumbered in both men and machines—and completely ignoring a standard rule of thumb, which states that an attacker must outnumber a defender by a ratio of at least 3:1 in order to ensure victory—the IDF never doubted that it would defeat the Egyptian army.

The IDF fought three break-in battles on the Sinai front. In the northern sector of the Sinai battlefield, an all-arms IDF division composed of tanks, mechanized and parachute infantry, artillery, and engineers punched through the Egyptian forward defenses in northern Sinai and Gaza.[14] It did so by engaging in traditional CAW at the tactical level. While tanks took the lead role in a number of tactical engagements, such as those in the vicinity of Khan Yunis and Rafah, they received substantial support from infantry, artillery, and engineers in these encounters. Parachute infantry took the lead role in other tactical engagements, in which tanks, mechanized infantry, artillery, and engineers assisted them. As a result of engaging in traditional CAW, the northern IDF division shredded two Egyptian infantry divisions, as well as their supporting tank and artillery forces, on the first day of the Six-Day War at modest cost to itself.

In one instance where this division did not practice traditional CAW—where tanks raced ahead of the infantry, artillery, and engineers—it suffered a rather bloody nose. Though a first wave of tanks passed unmolested through the Jiradi Pass on the road to the gateway to the Suez Canal in the northern Sinai, El Arish, because the Egyptian defenders had not been alerted to their presence in advance, a second wave of tanks was stopped in its tracks by the now alerted defenders. Not until combined arms teams of tanks, infantry, artillery, and engineers rooted out these defenders would the pass finally be cleared of the Egyptian adversary.

In the southern sector of the Sinai battlefield, another all-arms IDF division composed of tanks, parachute and other nonmechanized infantry, artillery, and engineers smashed the formidable Egyptian army defensive complex at Um Katef/Abu Ageila.[15] The IDF attack on this complex represents a classic example of traditional CAW at the tactical level. In a highly orchestrated night action, tank and infantry forces, under the cover of the heaviest artillery bombardment in IDF history to that date, struck the complex from the front, the flanks, and the rear. Engineers opened pathways through the extensive minefields that protected Um Katef/Abu Ageila so that tanks could provide direct fire support to the infantry forces assaulting the trench works around the complex. Parachute infantry, landed by helicopters in the Egyptian rear, knocked out the complex's artillery in a vertical flanking maneuver that took the Egyptian artillerymen completely by surprise. By the dawn of June 6, the fight for the complex at Um Katef/Abu Ageila was effectively over. The Egyptian infantry division garrisoning the complex, along with its associated tank and artillery forces, had collapsed under the Israeli blow, the remnants fleeing deeper into the Sinai.

In the central sector of the Sinai battlefield, an understrength IDF tank division composed of two armored brigades slogged its way over what the Egyptian army reckoned to be impassable terrain.[16] Elements of this division, emerging from a completely unexpected direction, thereby surprised elements of an Egyptian tank division racing forward at the same moment to bolster the Egyptian army's forward defensive positions in the northern Sinai. A tank-versus-tank engagement ensued at Bir Lafhan in which the superior quality of the Israeli crews, especially with respect to long-range gunnery, proved decisive. The Egyptian tanks were beaten back after suffering significant losses. IAF fighter-bombers contributed to the mauling of Egyptian forces in this sector, so the battle of Bir Lafhan represents an instance of non-traditional CAW based on armor and airpower, a pattern that would become more prevalent in the following days on the Sinai front.

By the morning of June 6, then, the IDF had won its three break-in battles. The Egyptian army's position in northeastern Sinai had become untenable, what remained of its forward defensive positions cut off from the rest of its forces and in the grip of utter chaos. Apparently in consequence of the drubbing its air force and frontline ground forces had received at the hands of the IDF, Egyptian general headquarters (GHQ) issued an order during the day for the rest of its ground forces to retreat back across the Suez Canal, even though much of the Egyptian army in the Sinai was still intact.[17] What was expected to be an organized withdrawal, however, quickly turned into a panic-stricken rout, as both officers and enlisted men abandoned their posts and equipment in great numbers in order to get back to Egypt proper by any means possible.

The Egyptian withdrawal order signaled the start of the pursuit phase of the IDF offensive in the Sinai.[18] IDF GHQ's prewar plan had been limited to Operation Focus and the break-in battles of Operation Red Sheet in northeastern Sinai; therefore, the pursuit phase of the Israeli offensive in Sinai was a rapidly improvised affair. IDF GHQ decided to employ a combination of armor and airpower to destroy as much of the Egyptian army in Sinai as possible before it could retreat across the Suez Canal. Armored spearheads from all three IDF divisions were to race westward and southward to secure the Mitla, Giddi, and Bir Gifgafa Passes—bypassing Egyptian strongpoints if possible, breaking through them if necessary—and thereby cut off the Egyptian army's routes of retreat. Tanks and fighter-bombers would then destroy the bottled up Egyptian forces in detail.

The pursuit phase of the IDF offensive unfolded essentially according to this blueprint, aided by the fact that the Egyptian army had lost its fighting spirit. Egyptian forces did occasionally put up stiff resistance, such as at Jebel Libni and the Bir Gifgafa Pass. IDF tanks, with a bit of assistance here and there from mechanized infantry and artillery, were able to overcome these forces in each instance. For the most part, however, Egyptian forces simply tried to avoid contact with IDF forces, which allowed Israeli armor to seize the Sinai passes and to shoot up Egyptian forces as they tried to make it through these escape routes. Concomitantly, IAF fighter-bombers, acting in loose coordination with the ground forces, wreaked havoc among Egyptian columns stalled on the Sinai's roads as a result of the IDF's control of the passes.

By June 9, the IDF had secured the whole of the Sinai, including the east bank of the Suez Canal, and the Six-Day War had come to an end on this front. The Egyptian army in the Sinai had lost 10,000–15,000 soldiers, as well as 80–85 percent of its military equipment.[19] Many hundreds of its tanks, APCs, and artillery tubes had been destroyed in the fighting, while still others had been captured intact by the IDF, eventually to be incorporated into the Israeli arsenal. The Egyptian army had also left behind some 10,000 assorted military vehicles, many of them captured intact as well. While IDF losses had certainly not been insignificant for an army of its size—losses had amounted to about 400 troops and perhaps as many as 120 tanks (some of which would later be repaired and reintroduced to service), they paled in comparison to those suffered by the Egyptian army. Rarely in the annals of military history had such a lopsided result been obtained in war.

Combined Arms Warfare on the Judean and Samarian Front

Israel sought to avoid war with Jordan in June 1967. The Israeli government relayed a message to this effect to the Jordanian government through United Nations (UN) and other diplomatic channels, emphasizing that the IDF would not act against Jordan if the Jordanian army did not act against Israel.[20] The Jordanian government, however, refused to heed the Israeli call to maintain the quiet between the two countries, committed as it was to the Arab war pact it had entered into days earlier and, perhaps, bamboozled by the Egyptian government's wild claims about nonexistent victories in the air and on the ground during the morning of June 5.

The Israeli government expected—and was prepared to exercise restraint in its response to—a show of force by Jordan in support of Egypt. An IDF offensive into Judea and Samaria would only be authorized if the scope, intensity, and duration of Jordanian attacks clearly exceeded the bounds of a show of force. The heavy shelling of Jewish neighborhoods in west Jerusalem, the occupation of UN-controlled territory in a contested area of the city, as well as air and artillery strikes against other targets within Israel, including vital IAF air bases, soon convinced the Israeli government that Jordan was not about to be content merely with a show of force, that it was indeed bent on an all-out war with Israel.

During the afternoon of June 5, therefore, the Israeli government authorized the IDF to launch an offensive into Judea and Samaria.[21] In consequence of Jordanian air and artillery strikes, the IAF had earlier been given the go-ahead to expand Operation Focus to include the Jordanian air force, which was destroyed by devastating attacks on its two main air bases. Now, with the government's approval, IDF GHQ authorized the ground forces' central and northern commands—IDF ground forces at the time of the Six-Day War were organized around three territorial commands: the southern, responsible for the Sinai front; the central, responsible for the Judean and Samarian front; and the northern, responsible for the Golan front—to commence Operation Whip, a contingency plan to crush Jordanian forces in Judea and Samaria and to push their remnants across the Jordan River. The IDF fielded about 40,000 soldiers and 200 tanks; its Jordanian adversary boasted approximately 45,000 soldiers and more than 250 tanks.[22]

The IDF launched a two-pronged offensive against the Jordanian army. In Judea, IDF Central Command unleashed a rapidly assembled, all-arms, division-sized battle group composed of tanks, mechanized and nonmechanized infantry (including parachute infantry), artillery, and engineers, some of its components drawn from GHQ reserves and IDF Southern Command, against three Jordanian infantry brigades, a tank brigade, and attached tank and artillery battalions. IDF ground forces employed traditional CAW at the tactical level in all of the battles in Judea, break-in and exploitation alike. Engineers opened paths through Jordanian minefields throughout Jerusalem and its environs for infantry and tanks to pass through. While parachute infantry played the lead role in defeating Jordanian forces in east Jerusalem, particularly in the Old City, it was supported throughout the fighting by armor, mechanized and nonmechanized infantry, artillery (sporadically), and engineers. On the northern and southern outskirts of Jerusalem, mechanized and nonmechanized infantry, tanks, and artillery functioned as combined arms task forces to overcome Jordanian infantry, tanks, and artillery. The same scenario unfolded in the fighting that occurred in the Jordan Valley and in the Hebron area.

In Samaria, IDF Northern Command deployed an all-arms division comprised of tanks, mechanized and nonmechanized infantry, artillery, and engineers against three Jordanian infantry brigades, an armored brigade, and attached tank, infantry, and artillery battalions. As in Judea, IDF ground forces waged traditional CAW at the tactical level in all of the battles in Samaria, including the large-scale engagements around Jenin and the Kabatiya Crossroads. Tanks, mechanized and nonmechanized infantry, artillery, and engineers functioned as combined arms task forces to overcome—sometimes with difficulty—their Jordanian adversary.

By the evening of June 7, the fighting had effectively come to an end in Judea and Samaria, the surviving Jordanian army forces retreating across the Jordan River. IDF ground forces had been significantly assisted in their swift offensive by IAF ground attack sorties, in an instance of CAW at the operational level. The Jordanian government's decision to order a general retreat of the Jordanian army on June 6 contributed as well to the Jordanian defeat. Though the Jordanian government later tried to rescind the order, the retreat proved impossible to stop once it had got underway.

The Jordanian army admitted to the loss of 700 soldiers, but this figure may represent a considerable undercount.[23] What is not in doubt is that it lost approximately 200 tanks, 50 APCs, and 200 artillery tubes, some of which were captured intact by Israeli forces and added to the Israeli arsenal. The Jordanian army also left behind thousands of other assorted military vehicles, some of them taken intact, too. The IDF lost approximately 300 soldiers and more than 100 tanks (at least some of which were repaired and returned to service). Its victory on this front had not only been costly, but also had not been nearly as lopsided as on the Sinai front.

Combined Arms Warfare on the Golan Front

Syria responded to the Israeli air and ground offensives against Egypt by staging air strikes and unleashing artillery bombardments throughout northern Israel—mainly against civilian targets—during the opening hours of the Six-Day War. The Israeli government answered these Syrian reprisals by permitting the IAF to extend Operation Focus yet again to include the destruction of the Syrian air force; however, it refused to sanction an IDF ground offensive to seize the Golan, even after the Syrian army launched a probing attack against Israeli territory on June 6—an attack that was easily beaten off, with significant losses inflicted on Syrian forces.[24] Despite the fact that Syrian artillery continued to pound Israeli targets, it would not be until June 9—that is, until after the conclusion of hostilities on the southern and central fronts—that the Israeli government would seriously consider a ground offensive on the northern front. The Israeli defense establishment did not relish the prospect of fighting on three fronts simultaneously. Until June 9, therefore, the IDF had to be content with air strikes and artillery bombardments against Syrian military targets on the Golan.

On June 9, under intense pressure from Israelis throughout the north of the country, who had been at the mercy of Syrian artillery for the past 19 years, the Israeli government abruptly reversed its course, authorizing the IDF to undertake Operation Hammer, the conquest of the Golan.[25] By this date, IDF Northern Command had amassed some 20,000 soldiers and 250 tanks, including those assigned from the southern and central commands by IDF GHQ.[26] The Syrian army deployed, perhaps, 50,000 soldiers and 500 tanks/tank destroyers on or in close proximity to the Golan, many of the troops hunkered down in extensive defensive fortifications. Moreover, the Syrian army held the high ground, as the Golan towered from 1,000–3,000 feet (305–915 m) above northern Israel.

As on the Judean and Samarian front, the IDF launched a two-pronged offensive. On the northern Golan, an all-arms division composed of tanks, mechanized and nonmechanized infantry, artillery, and engineers managed to ascend to the top of the Golan plateau and then break through Syrian army defensive positions, employing traditional CAW at the tactical level in both the break-in and exploitation battles. Under cover of air strikes and artillery fire, engineers led the assault, breaching the difficult terrain with armored bulldozers and opening paths through Syrian minefields. Once through the Syrian army's forward defensive positions, combined arms teams of tanks and infantry fanned out throughout the Golan, overcoming all Syrian opposition.

While the main IDF effort took place on the northern Golan, a second all-arms division moved into the southern Golan, primarily in order to confuse the Syrian army and divide its attention. By the morning of June 10, the IDF had made considerable progress in defeating the Syrian army on the Golan, despite occasionally meeting stiff resistance, as at the defensive complex around Tel Fakhr; however, Syrian forces still held much of the plateau.

Just as it seemed that Syria might hold onto part of the Golan, as a cease-fire loomed on the horizon, though, Syrian army GHQ ordered a general retreat from the plateau. Like the retreats carried out by the Egyptian and Jordanian armies, this one quickly turned into a rout, as Syrian troops simply abandoned their positions and equipment in a mad scramble to reach safety. The IDF took advantage of the Syrian retreat, converting Operation Hammer into Operation Pincers. In an effort to conquer as much of the Golan as possible in order to provide as much strategic depth on this front as possible, IAF helicopters lifted parachute infantry deep into the Syrian rear, where they eventually linked up with tank and infantry forces advancing from the west.

The Six-Day War came to an end during the early evening of June 10, with the IDF in control of the Golan. The Syrian army had lost some 500 soldiers (possibly many hundreds more) and some 120–160 tanks, about 40 of which fell into Israeli hands intact.[27] It also lost hundreds of artillery pieces, as well as hundreds of assorted military vehicles. IDF losses were put at approximately 125 soldiers and up to 160 tanks (though the latter figure is most likely to be a considerable exaggeration). In just over 30 hours, the IDF had overcome what many considered to be the Syrian army's impregnable defense layout on the Golan.

Combined Arms Warfare: Air–Ground Cooperation

In early June 1967, the Egyptian government knew that war between Egypt and Israel could erupt at any moment. Yet, Operation Focus, which the IAF had planned and practiced for years, caught the Egyptian air force completely by surprise on the morning of June 5.

Launching its aerial blitz at an unusual time—well after dawn—and from an unexpected direction—mainly from west to east—the IAF repeatedly struck 18 Egyptian air force bases in the Sinai and in Egypt proper with wave after wave of fighter-bombers.[28] Almost all of the IAF's approximately 200 frontline aircraft hit these targets in a pre-determined order that gave priority to those bases that housed the Egyptian air force's frontline fighter-bombers and long-range bombers. Continuous bombing and strafing runs were intended to disable runways, destroy aircraft, and pummel base infrastructure.[29]

Operation Focus dealt an immense blow to the Egyptian air force, which lost approximately three hundred aircraft, including most of its frontline fighter-bombers and all of its long-range bombers.[30] At least 13 air bases had been put entirely out of action, their runways cratered and their infrastructure pulverized. The other bases were not in much better shape. Numerous radar stations and surface-to-air missile (SAM) batteries had been destroyed as well. The Egyptian air force had essentially ceased to exist as a cohesive fighting force.

Much the same fate awaited the Jordanian and Syrian air forces (as well as one Iraqi air force base).[31] The IAF struck eight air force bases in these countries. Jordan lost its entire air force. Syria lost about half of its air force, including many of its frontline fighter-bombers, and Iraq lost a number of aircraft. The destruction of approximately 90 Syrian, Jordanian, and Iraqi air force fighter-bombers brought the total number of Arab aircraft eliminated during Operation Focus to about 400. The IAF lost 18 fighter-bombers to Arab air defenses (accounting for approximately 40 percent of its total losses in the war).[32]

Operation Focus had three principal objectives: (1) to eliminate the threat of large-scale air attacks on Israeli population and industrial centers; (2) to ensure that IDF ground forces would not come under sustained and effective air attacks; and (3) to ensure that the IAF would be free to provide sustained and effective air attacks on adversarial ground forces. Two of the operation's three objectives, in other words, were intended to provide assistance to IDF ground forces; therefore, despite the fact that Operation Focus was carried out exclusively by the IAF, it may be considered an instance of traditional CAW at the operational level, as airpower was acting in support of ground forces, even if indirectly so.

The IAF would fly a total of 2,591 close air support (CAS) and interdiction sorties throughout the Six-Day War.[33] Even though IDF ground forces had IAF air controllers embedded within them, most of these strikes were interdiction sorties. The IAF lacked an adequate command and control (C^2) infrastructure, as well as an adequate number of the right type of fighter-bomber (that is, low- and slow-flying, straight-winged aircraft), to conduct a high-tempo campaign of CAS strikes in immediate proximity to advancing IDF ground forces.

On the first day of the war, because it was occupied with the destruction of Arab airpower, the IAF flew a mere 268 CAS and interdiction sorties, 170 on the Sinai front and 95 on the Judean and Samarian front. Moreover, armed training aircraft flew many of these sorties. Hence, airpower had a very limited impact on the break-in battles on these fronts. On the following days, however, the IAF stepped up its ground attack sorties on both fronts. From June 6 through June 8, it flew 767 CAS and interdiction sorties on the Sinai front, while on June 6 and June 7, it flew 454 sorties on the Judean and Samarian front.

The IAF typically began by conducting strikes in the Egyptian and Jordanian armies' rear areas, its fighter-bombers gradually working their way back toward the ever-shifting front lines. While its aircraft were quite effective against "soft" targets, such as trucks and other thin-skinned vehicles, they did not destroy tanks or other armored vehicles in great numbers.[34] This inability to knock out tanks and other armored vehicles notwithstanding, the IAF still made two important contributions to the IDF ground campaign on each front.

First, its strikes sowed confusion and panic among Egyptian and Jordanian ranks, undermining the cohesion of their forces as they moved toward or away from the front lines, and often stalling or scattering those forces. Indeed, in numerous instances, Egyptian and Jordanian soldiers became so unnerved by IAF strikes that they simply abandoned their equipment en masse. Second, the destruction of large numbers of logistical vehicles clogged the roads over which the Egyptian and Jordanian armies had to travel, greatly impeding their forces' ability to advance or retreat. In consequence, IDF ground forces were usually able to outrun and/or outflank Egyptian and Jordanian forces in order to seize the most advantageous positions from which to destroy them. Though the IAF acted in rather loose coordination with IDF ground forces throughout the Six-Day War, its CAS and interdiction efforts on the southern and central fronts clearly played a part in the successful application of traditional CAW at the tactical level.

The IAF's CAS and interdiction effort on the northern front differed from its efforts on the southern and central fronts. For starters, the IAF carried out 433 sorties over a four-day period against Syrian army positions on the Golan prior to the commencement of the IDF ground offensive there. While these sorties did not result in a substantial amount of physical damage to the Syrian army—IAF munitions of the day could not penetrate Syrian fortifications—they did inflict a certain amount of psychological damage on the Syrian soldiers garrisoning them. Second, the IAF flew an additional 300 CAS and interdiction sorties in support of IDF ground forces during the break-in battle on June 9, a luxury not afforded to ground forces on either the southern or central fronts. Indeed, the IAF flew more ground attack sorties on the northern front—almost 1,100 of them—than it flew on either the southern or central front. Once again, its CAS and interdiction effort clearly played a role in the successful application of traditional CAW at the tactical level.

The IAF also contributed to the IDF's employment of traditional CAW at the tactical level in a logistical sense. On the Sinai front, it transported parachute infantry into the Egyptian army's rear on three occasions. It also delivered fuel to gasoline-starved tank forces racing to cut off the retreat of the Egyptian army.[35] And, on the Golan front, it transported parachute infantry into the Syrian army's rear.

Combined Arms Warfare: The Naval Component

The Six-Day War was nothing short of disastrous for the Israel Navy (IN). First, IDF GHQ abruptly canceled some of its operations, because the rapid pace of events had rendered them irrelevant to the Israeli war effort. Second, its special operations forces (SOF) failed in an attempt to sink Egyptian naval vessels at anchor in home waters. Third, and far more devastating, the IN spearheaded a joint naval and air attack on a United States Navy (USN) intelligence-gathering vessel that the navy mistook for an Egyptian warship, causing the deaths of more than 30 American sailors.[36]

The only "victory" registered by the IN in the Six-Day War occurred as a result of Operation Lights, a joint naval–land–air action to capture Egyptian army facilities at Sharm el-Sheikh, located at the southern tip of the Sinai.[37] The IN sent a task force of motor torpedo boats (MTBs) down the Red Sea to reconnoiter Egyptian defenses there and to provide fire support to the parachute infantry that was to seize the Egyptian facilities after being transported to the area by the IAF. An IN scouting party that went ashore, however, confirmed what aerial reconnaissance had discovered earlier—that the Egyptian garrison had bolted from the area. IDF GHQ, therefore, curtailed the navy's participation in the operation, electing to limit itself to sending in parachute infantry, who promptly overcame the feeble resistance offered by a group of Egyptian army stragglers and secured the facilities. For the IN, in short, the Six-Day War ended before it could make any meaningful contribution to the Israeli war effort.

Summary

The IDF engaged widely in traditional CAW throughout the Six-Day War. In the break-in battles on all three fronts, IDF ground forces employed tanks, mechanized and nonmechanized infantry, artillery, and engineers in coordinated, all-arms actions in order to pierce Arab defenses. These battles represent textbook cases of traditional CAW at the tactical level. The follow-on battles waged by IDF ground forces, particularly on the central and northern fronts, echoed the break-in battles in their reliance on traditional CAW at the tactical level. The IAF's opening strike on Arab air force bases, which essentially knocked Arab airpower out of the war, constitutes an example of traditional CAW at the operational level, albeit in an indirect sense. Much of the impetus behind Operation Focus, after all, revolved around providing assistance to IDF ground forces through the acquisition of air superiority over the battlefields. The emasculation of Arab airpower at the outset of the war not only ensured that IDF ground forces would not be subjected to effective and sustained air attacks, but also that those ground forces would reap the benefits of Israeli airpower as it ripped into Arab ground forces. The IDF's reliance on traditional CAW, in other words, accounts in large part for its spectacular victory in the Six-Day War.

Nevertheless, rather than endorse traditional CAW moving forward, the Israeli defense establishment chose to place its faith in a nontraditional version that privileged the tank–fighter-bomber combination at the expense of other arms, in consequence of the undeniable accomplishments of armor and aircraft throughout the Six-Day War, especially during the pursuit phase of the Israeli war effort on the Sinai front. No less an authority than the aforementioned American army expert on CAW concluded the following about the Israeli defense establishment's analysis of the war:

> The set-piece attacks conducted by teams of Israeli paratroops, infantry, artillery, and tanks to break open the Egyptian border defenses were forgotten in the euphoria over [an] armored dash to the Suez Canal. . . . The close and constant assistance of the Israeli Air Force made . . . artillery seem unimportant, especially in fluid operations when the Air Force [sic] could arrive more quickly than the artillery could deploy.[38]

The defense establishment, therefore, opted to double down on its bet that tanks and fighter-bombers would carry the day in future rounds of hostilities. This decision would later come to haunt the IDF.

3

The War of Attrition

Combined Arms Warfare Takes a Back Seat

In contrast to all of the other Arab–Israeli interstate wars (IWs), from which the State of Israel emerged with clear-cut victories, the 1969–70 War of Attrition came to a close in a stalemate. Egypt, on the one hand, did not "liberate" the Sinai, the principal objective of the Egyptian government in declaring the war in March of 1969. Conditions on the ground never became ripe for a crossing of the Suez Canal even to be seriously contemplated by the Egyptian army. Israel, on the other hand, did not bring about an end to the war on its preferred terms, the primary goal of the Israeli government during the hostilities. Indeed, the cease-fire agreement that terminated the War of Attrition left the country less secure than it had been prior to the outbreak of the war.

The manner in which the Israel Defense Forces (IDF) fought the War of Attrition differed fundamentally in two crucial respects from the manner in which it fought all of the country's other IWs. First, at least insofar as concerned the IDF's ground forces, the War of Attrition essentially remained a static contest that involved little in terms of movement. The ground forces' participation in the fighting consisted mainly of securing their forward positions along the Suez Canal and responding in kind to Egyptian army provocations. And, second, the Israel Air Force (IAF) shouldered most of the burden of prosecuting the war. The IAF would not again occupy such a central role in hostilities until the asymmetrical wars (AWs) of the twenty-first century, especially the 2006 Second Lebanon War.

The IDF's deviation from its usual interstate "way of war" during the War of Attrition resulted from the fact that the hostilities were never characterized by a "clash of armies." The IDF and the Egyptian army did not confront each other in large-scale, all-arms battles. The hostilities, to the contrary, were confined to artillery barrages, infantry raids, and air strikes across the canal, whose scope, intensity, and duration waxed and waned over a prolonged period of some 18 months.

Traditional (symmetrical) combined arms warfare (CAW) did not easily lend itself to this sort of fighting. Nor, for that matter, did nontraditional (asymmetrical) versions of the phenomenon. While largely absent from this war, however, traditional CAW did rear its head from time to time, especially at the tactical level, in respect of particular IDF actions during the fighting.

The Military Backdrop to the War of Attrition

The War of Attrition did not erupt spontaneously in early 1969. Rather, it grew out of the intermittent skirmishing between the Egyptian army and the IDF that began mere weeks after the conclusion of the 1967 Six-Day War. The War of Attrition, in other words, represented a greatly expanded—in scope, duration, and intensity—species of the fighting that raged along the Suez Canal from the summer of 1967 to the spring of 1969.

Hostilities between the Egyptian army and the IDF during this period typically unfolded according to a familiar pattern. The Egyptian army would attack IDF positions on the Israeli side of the canal, mainly employing artillery barrages but also occasionally infantry raids and air strikes to do so, with the intention of inflicting as many casualties as possible. The IDF, in turn, would invariably respond to these Egyptian provocations, usually by hitting Egyptian army positions (and, periodically, economic infrastructure targets, such as oil refineries) on the Egyptian side of the canal in like manner, but in a few instances striking deep inside Egyptian territory. In the autumn of 1968, to cite an example, IDF special operations forces (SOF) transported by IAF helicopters destroyed economic infrastructure targets—two bridges and a power transformer—in the Nile Valley.[1] Moreover, Israel also delivered thinly veiled threats to attack far more valuable economic infrastructure targets, such as the Aswan High Dam, by demonstratively conducting reconnaissance flights above them.[2] Israeli counterblows generally brought calm to the front, sometimes for months, but their deterrent effect would eventually wear off, leading to another round of tit-for-tat fighting.

Simultaneously, the Israeli defense establishment responded to the Egyptian army's attacks on IDF forward positions in the Sinai in a thoroughly uncharacteristic fashion for it at that time. The defense establishment decided to construct static defensive positions in the form of the Bar Lev Line (named after the IDF chief of staff on whose watch it arose), a string of very small and widely separated strongpoints along the Suez Canal, running from the Mediterranean Sea in the north to the Gulf of Suez in the south.[3] Mobile patrols consisting primarily of mechanized infantry secured the spaces between the strongpoints, while tanks and artillery tubes were positioned to the rear, ready to move up quickly in support of the strongpoints as necessary. The Bar Lev Line, it must be emphasized, was no Maginot Line in the Sinai. It was never intended to repel a full-scale, cross-canal invasion by the Egyptian army. Rather, its purpose was far more modest: to provide soldiers manning forward observation posts close by the Suez Canal with a measure of protection against Egyptian fire and to prevent Egyptian forces from infiltrating across the canal to establish a permanent presence in the Sinai. Construction of the Bar Lev Line, it must also be emphasized, did not signal a commitment by the Israeli defense establishment to move away from offensive maneuver warfare as the basis of the IDF's force structure and war-fighting doctrine.

A Summary of the Military Events of the War of Attrition

The War of Attrition proceeded through four distinct stages.[4] The first stage, which lasted from March 9–July 19, 1969, more or less followed the same pattern of artillery barrages and infantry raids that had characterized hostilities over the previous two years, but the fighting now was much more intense and long-lasting than in the past, leading to a spike in casualties on both sides of the Suez Canal. IDF artillery barrages and infantry raids no longer deterred the Egyptian army from continuing its attacks.[5]

Consequently, the Israeli government determined that Israel should "escalate [the war] for the sake of de-escalation" in the words of the contemporary IDF chief of staff.[6] The government reasoned that a massive air campaign against Egyptian positions along the canal would not only cut down on IDF casualties, but also would inflict sufficient pain on the Egyptian army to deter Egypt from continuing the war. Hence, for the first time since the end of the Six-Day War, the IAF entered the fray on a large scale, inaugurating the second stage of the War of Attrition, which lasted from July 20–December 31, 1969.

Up to July 20, the IAF's contribution to the fighting had been largely confined to defending the airspace over the Sinai against incursions by Egyptian air force fighter-bombers, to carrying out reconnaissance flights over Egyptian territory, and to transporting infantry forces behind Egyptian army lines. The IAF's limited participation in hostilities until this point in time may be gauged by the fact that the air force shot down only about 10 Egyptian fighter-bombers during this period.[7]

This state of affairs now changed abruptly. From mid-July until the end of December, the IAF would conduct an intensive bombing campaign against Egyptian army positions along the canal, acting as "flying artillery" in order to compensate for the IDF ground forces' gross inferiority in artillery tubes vis-à-vis its Egyptian adversary. The IAF's air campaign sought to eliminate the Egyptian army's integrated air defense system (IADS), which consisted of a mix of surface-to-air missile (SAM) and antiaircraft artillery (AAA) batteries (backed up by Egyptian fighter-bombers), as well as the Egyptian army's artillery array, which had inflicted the lion's share of the IDF's casualties in the fighting. The IADS had to be overcome in order to give the IAF a free hand against the artillery array.

By the close of 1969, the IAF had essentially eliminated the Egyptian army's IADS and had inflicted a substantial number of casualties among Egyptian troops.[8] It had also shot down another 35 or so Egyptian fighter-bombers.[9] Nevertheless, despite its control of the airspace over the canal zone, the IAF proved unable to suppress Egyptian artillery fire, not to mention mortar and small arms fire. The Soviet Union replaced the Egyptian army's artillery tubes faster than the IAF could destroy them. IDF casualties continued to mount at an unacceptable rate for a country with as small a population as Israel.

The Israeli government, therefore, decided to escalate the War of Attrition once again in order to attempt to end the hostilities on terms favorable to Israel. During the third stage of the War of Attrition, which lasted from January 1–April 18, 1970, in addition to keeping up the pressure against the Egyptian army along the Suez Canal, the IAF waged a "deep-penetration" bombing campaign against the Egyptian hinterland.[10] During these three-and-a-half months, the air force carried out 36 deep-penetration raids, encompassing a total of 118 sorties.[11] Most of the targets, all of them strictly military in nature (such as supply depots, training camps, SAM batteries, and radar stations), were deliberately chosen for their close proximity to Cairo and other population centers. On January 13, to cite one example, IAF fighter-bombers struck an Egyptian army base at Khanka, 12.5 miles (20 km) from Cairo. On January 28, to cite a second example, fighter-bombers hit another army base, this one at Ma'adi, a mere six miles (10km) from Cairo.[12] The Israeli government believed that the IAF's deep-penetration campaign would undermine the Egyptian government, thereby compelling Egypt to end the War of Attrition on terms favorable to Israel.

From a tactical standpoint, the deep-penetration campaign yielded impressive results. The IAF destroyed or damaged all of its targets—and without the loss of a single aircraft. From a strategic standpoint, however, the bombing campaign turned out to be a complete fiasco. First, the deep-penetration raids did not undermine the Egyptian government. In contrast to Israeli government expectations, the Egyptian populace rallied around its government, which had the effect of stiffening Egypt's resolve to continue the War of Attrition. And, second, the Soviet Union seized on the IAF raids as a pretext to insert itself directly into the fighting on behalf of its client. By the time that the IAF ceased its deep-penetration campaign, the Soviet army had already deployed between 6,500–8,000 troops and technicians to Egypt, including SAM crews and fighter-bomber pilots. This figure would grow to approximately 12,000 men, including no less than 120 pilots, by June of 1970.[13] Furthermore, IDF casualties continued to grow at an unacceptable rate.

The cessation of the IAF's deep-penetration campaign heralded the final stage of the War of Attrition, which lasted from April 19–August 8. Contrary to the second and third stages of war, when the initiative had been firmly in the IDF's hands, it now passed into the grasp of the combined Soviet–Egyptian forces, which were apparently emboldened to go over to the offensive by the Israeli government's decision to restrict IAF bombing to the canal zone. From mid-April until early August, these forces doggedly rolled forward an entirely new and far more capable IADS in some of the heaviest fighting of the War of Attrition in order to strip the IAF of its air superiority over the Suez Canal. The intensity of the hostilities during this stage of the war is indicated by the fact that the IAF shot down some 45 Egyptian fighter-bombers in the final months of fighting, including at least 5 flown by Soviet pilots.[14] Though the IAF inflicted severe manpower and matériel losses on Soviet–Egyptian forces as the latter moved ever closer to the Suez Canal, it could not ultimately prevent the completion of a sophisticated IADS along the canal. Moreover, the IAF suffered significant aircraft losses to the Soviet–Egyptian air defenses.[15] Indeed, by the time that a cease-fire agreement went into effect on August 8, the IAF had lost its air superiority in the airspace above and adjacent to the canal zone, a development that would contribute significantly to Israeli setbacks on the Sinai front at the outset of the next round of hostilities, the 1973 Yom Kippur War.

Not only had the War of Attrition ended quite unsatisfactorily from the Israeli point of view, but also it had been costly for the IDF.[16] From March of 1969 to August of 1970, the IDF lost over 250 men in the

Sinai or over Egypt. The number climbs to more than 350 when casualties from the summer of 1967 to the spring of 1969 are added to the total. That the Egyptian army suffered up to 10,000 dead over the same years did not offer any solace to the Israeli defense establishment. Nor did the far greater matériel losses incurred by the Egyptian army in comparison to those incurred by the IDF.

Combined Arms Warfare in the War of Attrition

While IDF artillery barrages and air strikes during the War of Attrition did not much lend themselves to CAW, the same cannot be said of its infantry raids. A number of these raids, many of which occurred during the second stage of the fighting, constitute classic examples of traditional CAW at the tactical level.[17] A closer look at the four most prominent of these raids is sufficient to demonstrate the validity of this assertion.

On the night of July 19/20, a joint Israel Navy (IN)–IDF ground forces team of special operations forces (SOF) personnel attacked and destroyed Egyptian army facilities on Green Island.[18] Employing a combination of bunker-busting munitions, machine guns, small arms, and explosives, the small Israeli team completely surprised and rapidly overcame a much larger force of defending Egyptian troops, blowing up the facilities on the island before withdrawing.

The next raid also involved a joint IDF naval–ground force team, though on a considerably larger scale. On the night of September 7/8, IN SOF personnel sank two Egyptian navy torpedo boats at anchor, eviscerating the Egyptian army's naval defenses in the Gulf of Suez and opening the way for a joint mechanized infantry–tank raid in the area.[19] On September 9, IN landing craft ferried a team of mechanized infantry and tanks to the Egyptian side of the Gulf of Suez. For 10 hours, this team advanced from one Egyptian army position to the next, with a special emphasis on radar installations, methodically destroying them, as well as inflicting heavy casualties in the process, before returning to Israeli territory aboard the same landing craft, having captured two of the latest model of Soviet tank to boot. IAF strikes assisted the raiders in their romp through Egyptian territory.

During the night of December 26/27, a joint IAF–parachute infantry team captured an advanced Soviet radar unit that the Egyptian army had set up in the Gulf of Suez area.[20] While IAF fighter-bombers engaged in diversionary air strikes to keep the Egyptian army's attention away from the raid site, air force technicians quickly disassembled the unit, which was then airlifted back to Israeli territory via helicopter. The fourth raid, this one against the Egyptian army garrison on Shadwan Island in the Gulf of Suez, was also a joint IAF–parachute infantry affair, though on a far larger scale.[21] A battalion-sized force of parachute infantry seized the island during the night of January 22, 1970, inflicting heavy casualties on the defenders. During their 30-hour stay on the island, the raiders removed a considerable amount of equipment, including radar units, again transporting their haul back to Israeli territory via helicopter. Air strikes that accompanied the raid sank another pair of Egyptian navy torpedo boats.

All of these raids could be thought of as having operational-level ramifications in regard to traditional CAW, as they assisted the IAF air campaign against the Egyptian IADS that began on July 20 in one way or another. Nevertheless, these raids cannot obscure the fact that, at the operational level, the IDF eschewed traditional CAW throughout the War of Attrition. A few years after the war, a former IAF commander would observe in this connection that:

> Instead of seeing the Air Force [sic] as one of the components in the fighting and putting in land forces and armor on a significant scale in order to crush the Egyptian land front with all its artillery and missiles, they [i.e., the Israeli defense establishment] entrusted the entire task to the Air Force [sic] alone and saw this as the sole solution, while the land forces continued to dig themselves in. What was needed was a land operation on a significant scale that would have [had] a wide and deep impact.[22]

The Israeli defense establishment thought that a large-scale, all-arms offensive on the Egyptian side of the Suez Canal would trigger a full-scale war, a development that it could not countenance. This fear, coupled with the defense establishment's growing infatuation in the years following the Six-Day War with a nontraditional version of CAW that privileged tanks and aircraft at the expense of other arms, lay behind the flawed decisions in July of 1969 and January of 1970 to rely on airpower to achieve an end to the War of Attrition on terms favorable to Israel. Airpower alone, the Israeli defense establishment blindly convinced itself, would be sufficient to defeat the Egyptian adversary.

These comments, of course, are not meant to imply that the war would necessarily have concluded on much more favorable terms had the IDF's actions at the operational level been based on the principles of traditional CAW. An argument can certainly be made that the results would have been better for Israel had the IDF engaged in traditional

CAW at this level, but what would actually have happened had this scenario played itself out must remain in the realm of speculation.

Summary

The Israeli defense establishment considered the War of Attrition to be an aberration. The next round of interstate hostilities, it averred, would revert to the pattern set by the 1956 Sinai Campaign and the Six-Day War—that is, the next interstate conflagration would be a war of maneuver, not a war of attrition. The defense establishment further determined that, precisely because the War of Attrition was an aberration, the IDF's experience in the hostilities was sui generis and, therefore, did not merit a rethink of the IDF's post-Six-Day War force structure and war-fighting doctrine, especially at the operational level. The defense establishment, in short, rather peremptorily dismissed the suggestion that the IDF could derive lessons from its experience in the War of Attrition that would be applicable to a future war.

In respect of its first conclusion, the Israeli defense establishment got it right. The Yom Kippur War did indeed turn out to be a war of maneuver, certainly on the Israeli side of the ledger. In respect of its second conclusion, however, the defense establishment got it wrong. The belief that the IDF's experience in the War of Attrition held no relevance for the next round of hostilities was quickly discredited at the outset of the Yom Kippur War, particularly in regard to the IAF's initial encounters with the Egyptian and Syrian armies' IADSs. The defense establishment, to be sure, had acknowledged the air force's bruising encounter with Egyptian air defenses during the last stage of the War of Attrition, but it firmly adhered to the notion that early warning of the next war, which it blithely assumed would be provided by the Israeli intelligence community, coupled with elaborate IAF plans to strike adversarial air defenses at the start of the war, would be sufficient to address the problem. Even the failure to inflict any damage whatsoever on the Egyptian IADS during an unopposed 1971 IAF strike against the system in response to Egyptian army provocations in the Suez Canal zone—a strike that utilized some of the most sophisticated American-supplied antiradiation missiles (ARMs) and electronic countermeasures (ECM)—did not rouse the defense establishment from its complacency.[23]

Again, these comments are not meant to imply that the dangerous reverses suffered by the IDF during the opening days of the Yom Kippur War would necessarily have been avoided entirely had the Israeli defense establishment properly digested the lessons of the War of Attrition; however, had the defense establishment modified the IDF's force structure and war-fighting doctrine in a manner compatible with the principles of traditional CAW instead of wholeheartedly buying into a nontraditional version of the phenomenon based on tanks and aircraft, it is at least plausible to contend that the IDF would have fared better than it actually did during the early days of the war.

4

The Yom Kippur War

Combined Arms Warfare Takes a Licking but Keeps on Ticking

The Israel Defense Forces (IDF) emerged from the 1973 Yom Kippur War as the undisputed victor over the armies of its Arab foes. Despite suffering serious reverses on both the Golan and Sinai fronts during the opening days of the war, the IDF not only recovered quickly to inflict on Arab armies losses in men and machines that far exceeded its own, but also it captured considerably more territory than it had relinquished during the fighting. In the north, the IDF first swept the Syrian army entirely from the Israeli-controlled portion of the Golan and then proceeded to conquer a swath of territory inside Syria. In the south, though the Egyptian army remained ensconced in a strip of territory in the Sinai throughout the war, the IDF eventually crossed the Suez Canal and seized a sizable chunk of land inside Egypt proper. By the war's end, the IDF had advanced to within striking range of Damascus and Cairo, the Syrian and Egyptian capitals, respectively.

Nevertheless, the IDF came under intense and widespread criticism for its performance during the Yom Kippur War. Perhaps the most frequent criticism leveled against the IDF asserts that it shunned the practice of traditional (symmetrical) combined arms warfare (CAW) in favor of a nontraditional (asymmetrical) force structure and war-fighting doctrine that relied overwhelmingly on tanks and aircraft at the expense of infantry, artillery, engineers, and logistics. The following are three rather typical examples of this particular criticism:

> [The IDF's] attempts to use tanks and air power [sic] without strong support from the infantry or artillery branches . . . led to many of its problems during the 1973 war and showed [it] the need for a more balanced combined arms approach.[1]

> Both the [Sinai Campaign] and the Six-Day War had left the [IDF] with the impression that wars on the ground were won by armor [i.e., tanks] and armor alone. As a result, [it] failed to develop an integrated infantry–armor doctrine, and effectively eschewed the use of infantry. . . . Part of the IDF's problem was its over-reliance on armor; another equally important component was its underreliance on artillery. The latter was related to the fact that . . . the [Israel Air Force] had gained air superiority within the first few days of [previous] conflict[s].[2]

> The IDF of 1973 was not built organizationally or mentally for an absolute commitment to joint operations. The realization of the paramount value of combined-arms [sic] combat was one of the IDF's bitterest lessons of the Yom Kippur War. . .[3]

Many more examples could be added to the list, of course, but these three—the first two by American military analysts, the third by an Israeli tank battalion commander on the Sinai front—suffice to reveal the essence of the criticism with respect to the IDF's force structure and war-fighting doctrine during the war.

This criticism, the truth be known, certainly has much merit behind it. As a consequence of its previous experience, particularly during the 1967 Six-Day War, the IDF built up its tank and aircraft order of battle (ORBAT) at the expense of its infantry, artillery, engineer, and logistics ORBAT in the years prior to the Yom Kippur War. Furthermore, the IDF's war-fighting doctrine before the war at both the tactical and operational levels of warfare assigned decisive roles to tanks and aircraft, foregoing a more traditional approach to the integration of combat arms.[4]

Be that as it may, this criticism is based primarily on the IDF's performance during the first few days of the Yom Kippur War, especially on the Sinai front, when it scrambled to block the Syrian and Egyptian assaults with the meager forces that were deployed in the north and south at the commencement of hostilities. Indeed, many of the IDF's critics concede that its shortcomings in CAW were the result not only of its faulty force structure and war-fighting doctrine before the war, but also of the circumstances that it faced at the outbreak of hostilities. Once the IDF recovered from its initial shock and stabilized the fronts, its performance in respect of CAW displayed marked improvement on both the tactical and operational levels.

Combined Arms Warfare in the IDF Before the Yom Kippur War

The Israeli defense establishment concluded that the country's tremendous victory in the 1967 Six-Day War should be attributed almost exclusively to the actions of the Israel Air Force (IAF) and the IDF tank corps. This view had serious consequences for the IDF's force structure and war-fighting doctrine. Already favored in the IDF's procurement plans well before 1967, tanks and aircraft received even more emphasis after the war. The IDF's tank inventory increased by roughly 100 percent and its fighter-bomber inventory by more than 50 percent between 1967 and 1973, not to mention that their quality improved dramatically as well with the acquisition of the most modern American weapons. Indeed, the IAF's acquisition of state-of-the-art American fighter-bombers, which replaced most of its antiquated French aircraft, increased its striking power several times over.[5] Outlays on the IAF alone ate up over 50 percent of the Israeli defense budget in the years prior to the Yom Kippur War.[6]

Other IDF branches had to make do with what was left of the budget after tank corps and air force needs were addressed by the defense establishment. The capabilities of these branches, therefore, either remained stagnant or withered over time. The mechanized infantry went to war in 1973 with only a limited number of modern armored personnel carriers (APCs), and most of both the mechanized and nonmechanized infantry entered hostilities with antiquated rifles and antitank weapons. Furthermore, the quality of the mechanized infantry's manpower had deteriorated during the interwar period.[7] The artillery branch had roughly the same number of tubes in 1973 as it had in 1967, even if their overall quality had improved somewhat with the acquisition of small numbers of modern American self-propelled guns.[8] The engineering branch did not possess enough vehicles to move its equipment around the battlefield, nor did the logistics branch possess enough tank transporters to move quickly the required number of tanks to the battlefield. And many of those tanks were not ready for battle on the eve of hostilities, because the IDF's logistics branch lacked sufficient resources.[9]

The IDF's war-fighting doctrine evolved along a parallel track during the years from 1967 to 1973. The Israeli defense establishment concluded that tanks—whether the IDF was on the offensive or the defensive, whether it was engaged in a maneuver or an attrition campaign—did not need much in the way of infantry or artillery support. Mechanized and nonmechanized infantry would be employed mainly to mop up and secure the battlefield after the tanks had won the day. Suppressive fire from artillery would not be necessary on a large scale, because the defense establishment was convinced that the IAF's fighter-bombers would act as "flying artillery" in support of the tanks as soon as air superiority had been won—and the defense establishment took it for granted that the IAF would quickly achieve air superiority in the next war. The roles of the engineering and logistics branches in the IDF's war-fighting doctrine received no in-depth consideration from the defense establishment from 1967 to 1973.

This nontraditional vision of CAW was itself based on a number of assumptions, collectively termed "the concept," held by the defense establishment about the circumstances under which the next war would break out.[10] A first assumption held that the Arabs would not initiate a war that they could not win. A second assumption held that the Arabs would not begin a war until they could neutralize Israeli airpower by striking at IAF bases in Israel proper. And a third assumption held that Israeli military intelligence, A'MAN, would in any case provide at least 48–72 hours of advanced warning of any attack.

All three of these assumptions turned out to be erroneous. The Arabs were willing to suffer a battlefield defeat in order to achieve a diplomatic victory; they were willing to sacrifice their armies in order to spark a diplomatic process that they believed would lead to the return of the Golan and Sinai. The Arabs had also devised a sophisticated operational plan to neutralize Israeli airpower that did not involve strikes against IAF bases. Instead, they would construct highly sophisticated integrated air defense systems (IADS) composed of large numbers of surface-to-air missile (SAM) and antiaircraft artillery (AAA) batteries on both fronts to blunt the effects of Israeli airpower. And, lastly, A'MAN would not in the event provide a clear war warning 48–72 hours in advance of the Arab attack. Though it monitored the deployment of the Egyptian and Syrian armies along the cease-fire lines for weeks before the attack, and though it informed both IDF general headquarters (GHQ) and the Israeli government that the Arabs had the ability to launch an attack given their present deployments along the fronts several days in advance of the attack, it continued to insist that the probability of an Arab attack in the fall of 1973 was quite low.

For their part, both IDF GHQ and the Israeli government, like A'MAN, chose to focus on Arab intentions rather than capabilities. Hence, they did not authorize a full-scale mobilization and deployment of the IDF's reserve formations, which constituted the bulk of Israel's fighting power, until literally a few hours before the commencement of hostilities. Not until an unimpeachable intelligence source had unequivocally confirmed that Egypt and Syria intended to launch an attack very shortly did Israel even contemplate this step.

Consequently, at approximately 14:00 hours on October 6, when the Syrian and Egyptian armies opened fire simultaneously on the Golan and in the Sinai, respectively, the IDF faced grossly lopsided force ratios on both fronts.[11] On the Golan, the Syrians had a roughly 5:1 advantage in tanks (an 8:1 advantage counting their second-echelon tanks), a 10:1 advantage in infantry, and a 12–13:1 advantage in artillery tubes along the forward line of contact.[12] In the Sinai, the Egyptians had a roughly 14:1 advantage in tanks, an 80:1 advantage in infantry, and a 40:1 advantage in artillery tubes along the forward line of contact.[13] The IDF's skewed force structure and war-fighting doctrine, combined with its lack of readiness for the coming war, would result in disastrous consequences for it during the opening days of the war.

Combined Arms Warfare on the Golan Front

The IDF's war on the northern front can be divided into three distinctive phases: a defensive phase from October 6 through October 10 to clear the Golan of Syrian forces, an offensive phase from October 11 through October 14 to acquire additional territory in Syria proper, and a second defensive phase from October 15 through October 24, when a stable cease-fire came into effect, to hold onto this newly captured territory. At the outset of hostilities, the IDF had the better part of two high-quality tank brigades (the southern of the two was less one tank battalion), a series of small hilltop strongpoints manned by high-quality infantry, a handful of artillery batteries, and some mechanized infantry and engineering forces to meet the Syrian onslaught of three infantry divisions, three tank brigades, more than 100 artillery batteries, and assorted engineering forces (with two additional tank divisions close by in the rear).[14]

The Syrian army launched a two-pronged offensive across the 1967 cease-fire line. Its main thrust was aimed at the southern part of the Golan, with a more limited (but still very potent) thrust aimed at the northern part of the plateau.[15] IDF ground forces on the Golan were better prepared to wage a defensive struggle against an invader than their counterparts in the Sinai. The high-quality infantrymen on the front line not only defended themselves and harassed Syrian columns that bypassed them, but also they served as forward observers for artillery barrages and air strikes against rear-echelon targets.[16] Some of the strongpoints held out until relieved later in the fighting, while those in danger of being overwhelmed by the Syrian army were evacuated in an orderly manner. Only one would fall to the Syrian army.

On the northern Golan, the Syrian offensive got nowhere. A spectacular mobile defense waged by the IDF tank brigade stationed in the area stopped much larger Syrian ground forces in their tracks for several days, destroying hundreds of tanks and APCs in the process. On the southern Golan, to the contrary, the Syrians were able to break through the IDF tank brigade in this sector. Though this brigade took a very heavy toll of attacking Syrian tanks and APCs, the gross imbalance in numbers between attacker and defender proved decisive here. A breakthrough into Israel proper—which had now become a very real possibility, even if was not part of Syria's prewar operational plan—was only prevented by the delaying actions of the remnants of this brigade, IAF air strikes against rear-echelon supply convoys, and the timely intervention of the first reserve forces, which arrived within 12 hours of the outbreak of hostilities, as well as Syrian hesitance.[17]

By October 8, the IDF had amassed enough forces on the Golan—significant elements of three tank divisions—to begin a counterattack against the Syrian army in order to push it back beyond the "Purple Line," the title given to the 1967 cease-fire line. This multidivisional effort on the southern Golan steadily drove the Syrian army out of Israeli-controlled territory. Indeed by October 10, the Syrian army held only one position inside the Israeli portion of the Golan, a strongpoint on Mount Hermon, which the IDF would recapture late in the war.

Though tanks spearheaded the defensive effort from October 6 through October 10, the IDF did practice traditional CAW to a certain extent during the opening days of the war on the Golan front. At the tactical level, the hilltop strongpoints acted as forward observation posts in directing artillery barrages and air strikes against targets in the Syrian army's rear. Artillery engaged in suppressive fire in support of IDF tank forces whenever possible, while engineering forces created or demolished obstacles to slow the Syrian advance.[18] At the operational level, the tanks, mechanized and nonmechanized infantry, artillery, and engineers of three divisions engaged in a coordinated effort to drive the Syrian army from Israeli-controlled territory. And air support, though acting only in loose coordination with the ground forces, proved important in halting the Syrian advance on the southern Golan.[19]

On October 11, the IDF began a counteroffensive into Syria.[20] The principal purpose of this attack was to capture a swath of territory inside Syria to act as a bargaining chip in postwar negotiations, particularly in light of the fact that Egypt had managed to conquer a chunk of territory in the Sinai. Traditional CAW played a prominent role in this counteroffensive, as the IDF "advanced in a manner designed to minimize casualties by using heavy tank and artillery fire to open the way rather than making costly armor charges."[21] Mechanized and parachute infantry, engineering forces, and IAF fighter-bombers also performed prominent tasks in the IDF's attack. On October 14, the IDF halted its counteroffensive, when its long-range artillery had moved to within range of the outskirts of Damascus.

From October 15 through October 24, the Syrian army, along with Iraqi and Jordanian expeditionary forces, engaged in local counterattacks against the Israeli enclave. The IDF repulsed all of these attacks through the practice of traditional CAW. Tank, mechanized and parachute infantry, artillery, and engineering forces worked in unison to beat back assaults by Arab tank and infantry forces. The only Israeli-initiated attack during the last 10 days of fighting on the Golan front involved the recapture of the IDF intelligence-gathering facility on Mount Hermon.[22] Infantry forces advancing from above and below, with artillery and air support, won a bloody triumph over Syrian special operations forces (SOF) ensconced there in a feat of traditional CAW at the tactical level.

Combined Arms Warfare on the Sinai Front

The IDF's war on the southern front can also be divided into distinct phases: a largely defensive phase from October 6 through October 14 to halt the Egyptian penetration in the Sinai and an offensive phase from October 15 through October 24 to cross the Suez Canal in order to seize territory in Egypt proper. When the Egyptian army initiated its assault on Sinai, the IDF had only a few hundred, low-quality infantrymen, a few dozen tanks, and a handful of artillery tubes immediately adjacent to the waterline to oppose the five infantry divisions of the first wave, which included hundreds of tanks and APCs and which were supported by about 2,000 artillery tubes.[23]

The IDF's infantrymen were holed up in the Bar Lev Line, a series of widely scattered strongpoints alongside the Suez Canal. Though this line was not built to withstand a large-scale, cross-canal attack, the Israeli government demanded that the IDF prevent an Egyptian "land grab" in the Sinai that could result in a diplomatic setback for Israel. Thus, the Israeli defense establishment, rather than permit the infantrymen in the strongpoints to withdraw to safety in case of a large-scale assault, decided that they must remain in place in order to harass the invading Egyptians and to act as forward observers for artillery fire and air support until they could be relieved by an IDF counterattack.

This forward defense concept, combined with the Israeli ethos of never abandoning men in the field, created a harrowing scenario for the IDF at the outset of the war. The low quality of the infantrymen on the Bar Lev Line, as well as the immense volume of firepower thrown at them by the Egyptian army, meant that the strongpoints could neither effectively harass the infantry divisions crossing the canal nor effectively call in artillery barrages or air strikes (even had artillery or air support been a realistic option). Moreover, with many of them surrounded and under siege, the strongpoints began to call for assistance.

The IDF responded with small-scale "tank charges" toward the canal in an effort to relieve the strongpoints. These tank charges were not supported by artillery fire, as the few guns near the front were under heavy counterbattery fire themselves, or by infantry, as most of the infantry in Sinai had been drained off either to protect rear-area installations or to counter Egyptian SOF troops transported into the Israeli rear via helicopters, or by fighter-bombers, as the IAF's squadrons were busy with other tasks. Attacking into the teeth of extremely heavy and coordinated artillery, tank, and antitank missile and rocket fire without any supporting arms, the three tank brigades committed to battle on that first day lost approximately 150 tanks (closer to 175 if mechanical breakdowns and accidents are counted as well), with precious little to show for their efforts.[24] Though tanks did manage to link up with a few of the strongpoints and to evacuate some of the wounded, all of the Bar Lev Line positions, except for one in the extreme north, along the Mediterranean coast, eventually fell to the Egyptian army, with most of the infantrymen in them either killed or captured.

While the IDF's regular tank division was fighting to relieve the strongpoints and to contain the Egyptian bridgeheads along the canal, two reserve tank divisions arrived in Sinai in less than two days, despite a lack of tank transporters and other logistical problems. The mobilization and deployment of tanks took priority over infantry and artillery. Once IDF GHQ believed that a sufficient number of tanks had been accumulated along the front, it authorized a multidivisional counterattack on October 8. The IDF, after all, had been built primarily for offensive maneuver warfare, and its war-fighting doctrine emphasized a counterattack at the earliest possible moment.

This counterattack, averred GHQ, would roll up the Egyptian bridgeheads from north to south and, perhaps, capture a bridge, which the IDF could then use to cross the canal into Egypt proper.[25] This optimistic assessment proved to be divorced from reality. The counterattack quickly turned into a debacle. Though the IDF probably gave as good as it got from the Egyptian army, two of its tank battalions were more or less completely destroyed during the fighting, in part by infantry equipped with antitank missiles and rockets. When the IDF finally halted its counterattack, the Egyptians had not been dislodged from any of their positions in the Sinai.

The counterattack violated the most basic principles of traditional CAW at both the tactical and operational levels: symmetry and integration. Infantry, artillery, and air support were sparse at best. The lead division in the attack, for example, had been stripped of its mechanized infantry and had only a handful of artillery tubes to engage in suppressive fire.[26] Despite the battering that the IDF's tanks had absorbed on October 6–7, they once again charged into the teeth of Egyptian defenses on their own, with predictable results.

On the Golan front, the lack of infantry and artillery forces did not constitute a major problem, because the challenge faced by the IDF came mainly from Syrian tanks. The hilly, rocky terrain made it difficult for the Syrian army to deploy antitank missiles and rockets, so these weapons played almost no role here. In the Sinai, with its largely flat terrain, to the contrary, antitank missiles and rockets had a very prominent role in Egypt's operational plan. The paucity of infantry and artillery during the opening days of the war, therefore, had grave consequences for the IDF's fortunes here, as infantry and artillery were the most effective means for countering these weapons.

Command and control (C^2) problems throughout the October 8 counterattack added to the IDF's woes. The two tank divisions involved in the fighting did not coordinate their operations at any point in the battle, and one of them essentially went around in circles for much of the battle as a result of confused and contradictory orders issued by IDF GHQ and IDF southern command headquarters (HQ).[27]

Bruised by its setback in the counterattack, the IDF went back on the defensive from October 9 through October 14. During this period, its force structure and war-fighting doctrine accommodated themselves to the lessons of the first three days of battle: sufficient numbers of infantrymen and artillery tubes were incorporated into the tank divisions. Damaged tanks were repaired and returned to service. And, from this point forward, the IDF would no longer engage in unsupported tank charges. Rather, tanks would advance cautiously, and in tandem with suppressive fire from infantry and artillery, mindful of the advantages of traditional CAW.[28]

From October 9 through October 13, both the IDF and the Egyptian army were largely content to hold and consolidate their positions, satisfying themselves with limited probes along the front line. The skirmishes that erupted as a consequence of these probes routinely ended in the IDF's favor, because of its clear superiority in maneuver warfare.[29] During this period, IDF GHQ decided to wait for the Egyptian army to launch a large-scale offensive in an attempt to break out of its canal-side bridgeheads, which GHQ expected to come shortly, before initiating a cross-canal offensive of the IDF's own.

The opportunity to engage in a major mobile tank battle occurred on October 14. Partly in order to reach the key Sinai passes and partly in order to reduce the pressure on its faltering Syrian ally in the north, the Egyptian army launched a multidivisional offensive along a number of different axes. To make a long story short, the Egyptian army suffered a crushing defeat that day, losing 200–250 tanks. IDF tank losses stood at 10–20.[30] At the tactical level, suppressive fire from IDF infantry and artillery countered the Egyptian infantry's antitank missiles and rockets, turning them into a nonentity during the battle, while Israeli tanks dealt harsh blows to their Egyptian counterparts. At the operational level, IAF fighter-bombers cooperated with IDF tanks, infantry, and artillery to halt and scatter any Egyptian forces that moved beyond the range of the Egyptian army's IADS. In the October 14 tank battle, the largest in history, except for the Battle of Kursk during the Second World War, the IDF performed a genuine feat of traditional CAW.

If the October 14 tank battle constituted the first turning point of the war on the southern front, the IDF's cross-canal counterattack, begun on October 15, constituted the second turning point. From this date until the end of the war, the IDF would be on the offensive, the Egyptian army on the defensive. The crossing operation itself, known as Operation Strongheart (or, alternatively, as Operation Stout-hearted Men), was another feat of traditional CAW at both the tactical and operational levels. Tanks, mechanized and nonmechanized infantry, artillery, and engineers all worked hand in hand, first, to break through to the waterline and, second, to ferry tank and infantry forces across the canal.[31] Essentially, IDF tank brigades, infantry battalions, and artillery battalions fought a bloody but ultimately victorious battle, particularly at a place called the Chinese Farm (which had actually been a Japanese agricultural station prior to the Six-Day War), to break through to the waterline. Meanwhile, engineers ferried parachute infantry and a number of tanks across the canal on boats and pontoons, creating a small bridgehead on the Egyptian side of the canal. After several days of additional hard fighting, the IDF expanded and solidified its bridgehead on both sides of the canal, eventually transferring the better part of three tank divisions into Egypt proper over bridges that its engineers had constructed across the canal.

Once across the Suez Canal, mixed tank and infantry columns immediately began to fan out, especially to smash SAM and AAA batteries.[32] IDF artillery fire from the Israeli side of the canal also hit batteries located close to the waterline. These efforts resulted in the destruction of wide swaths of the Egyptian army's IADS, opening up undefended flight corridors, which in turn allowed IAF fighter-bombers to provide air support to advancing IDF columns. By the time a more or less stable cease-fire came into effect on the southern front, the IDF had advanced to within about 60 miles (100 km) of Cairo, had surrounded the Egyptian Third Army (one of the two corps involved in the Egyptian offensive), and had captured more territory on the Egyptian side of the canal than Egypt held on the Israeli side. The practice of traditional CAW at both the tactical and operational levels had once again proven itself.

Combined Arms Warfare: Air–Ground Cooperation

In the months prior to the Yom Kippur War, the IAF concluded that it would need 48 hours to take command of the air at the outset of hostilities if given a clear war warning 36–48 hours ahead of time.[33] The air force preferred to launch a preemptive strike to achieve this objective, but it asserted that it could still accomplish the objective if the Arabs attacked first. The IAF's main priorities at the start of hostilities would be to ensure air supremacy over Israel itself, so that the IDF's reserve formations could mobilize and deploy for battle unmolested by Arab airpower, and to achieve air superiority over the battlefield(s), so that it could then provide air support as needed to IDF ground forces. To accomplish these two tasks, the IAF developed intricate operational plans to demolish the Syrian and Egyptian IADSs—operational plans that included cooperation with IDF long-range artillery—as well as to pin down the Syrian and Egyptian air forces through a combination of air-to-air combat and air base attacks.

The IAF's prewar thinking assigned a rather low priority to support of the ground forces. Indeed, its intent to take command of the air at the outset of hostilities meant that the ground forces would be largely, if not entirely, devoid of air support during the first two days of war. The IDF's ground forces, reasoned both the IAF and IDF GHQ, would be able to hold their positions against the Arab armies until the air force could join the battle in a decisive way. The IAF had developed an operational plan to deal with a worst-case scenario in which IDF ground forces required immediate, large-scale air support to hold their positions, but it sat in a file cabinet somewhere gathering dust; the air force never gave it much consideration.

Despite the fact that Israel had only a few hours of warning that war was about to erupt in the north and south, the IAF still possessed the capability to launch a preemptive strike, albeit one more modest in scope than the one envisaged in its elaborate prewar operational plans. Be that as it may, the Israeli government vetoed a preemptive strike for two major reasons. First, and foremost, it feared the loss of American military and diplomatic support if the IDF opened fire before the Arab armies.[34] And, second, IDF GHQ had assured the government that Israel would win the war even if the Syrian and Egyptian armies struck the initial blow.

Opinions are mixed as to the impact an IAF preemptive strike might have had on the IDF's fortunes in the war. On the one hand, some military analysts believe that even a limited strike would have completely changed the complexion of the war by shifting the operational initiative to the IDF. Two analysts, for example, calculated that, had Israel seized the initiative, the IAF could have knocked out 90 percent of the Syrian and Egyptian SAM batteries "in a period of three to six hours for the loss of under [10] aircraft." With air superiority over the Golan and Sinai thus virtually assured as a consequence, according to this line of reasoning, IAF intervention in the ground battle would have been swift and massive, its fighter-bombers dropping "[3,000] tons of bombs on enemy targets before the Arab attack reached full strength."[35] Other analysts, in contrast, believe that a preemptive strike, regardless of its scope, would not have made an appreciable difference in how the war unfolded in the first few days.[36] Which perspective has greater merit, of course, must remain in the realm of speculation.

Regardless of the fact that the war began under the worst-case scenario, the IAF devoted the lion's share of its resources during the opening days to its prewar priorities: securing Israeli airspace and overcoming Arab IADSs. On the morning of October 7, the IAF initiated a large-scale attack against the Egyptian IADS based on its prewar operational plan. The attack initially made headway, but it was soon canceled, because of the ominous situation on the Golan front.[37] The IAF then attempted to attack the Syrian IADS, again based on its prewar operational plan. This attack, which resulted in heavy aircraft losses as a result of outdated intelligence information and uncoordinated execution, failed to dent, let alone destroy, Syrian air defenses.[38] Hence, the Syrian IADS survived the war mainly intact and the Egyptian IADS was only overcome late in the fighting, largely as a result of the actions of IDF ground forces.

The inability to knock out the Syrian and Egyptian IADSs early on in the war seriously impeded the IAF's ability to engage in traditional CAW with the ground forces throughout much of the war.[39] On the Golan front, in order to avert a potential Syrian breakthrough into Israel proper, the IAF did act as flying artillery out of extremely dire necessity during the first few days of the war, engaging in a high-tempo campaign against the Syrian army. Its strikes against rear-echelon, soft-skinned supply convoys were helpful in slowing the momentum of the Syrian offensive, but at a very considerable cost in aircraft lost to air defenses. After the IDF advanced into Syria, the IAF's efforts on the Golan slackened considerably in order to conserve the air force's fighter-bomber fleet.

During the first days of war, IDF ground forces on the Sinai front received less effective air support than those on the Golan front. The support the IAF furnished here was aimed primarily at countering the infiltration of Egyptian SOF, at relieving the Bar Lev Line strongpoints, and at knocking out the Egyptian bridges. Cooperation between the IAF and IDF ground forces succeeded in eliminating the SOF threat, as the air force shot down many troop-carrying helicopters and ferried IDF "commando-hunter" squads to deal with those Egyptians who managed to land behind the front lines. In both of the latter cases, however, not only did air support prove very ineffective, but also it entailed significant losses to the Egyptian IADS, which is why these efforts were soon suspended. The IDF counterattack on October 8, as noted above, received little air support.[40] After the first few days of hostilities, the IAF limited its support to the ground forces almost exclusively to areas that fell outside of the Egyptian IADS, as in the October 14 tank battle. Only after IDF ground forces crossed the canal and began to overrun Egyptian SAM batteries did the IAF really enter the land war in a more significant manner, helping to seal the fate of the Egyptian Third Army.

Overall, the IAF and IDF ground forces did not act as an efficient CAW team at either the tactical or operational levels throughout much of the war, a few instances to the contrary notwithstanding. The inability to overcome the Syrian and Egyptian IADSs at the outset of hostilities certainly goes a long way toward explaining the deficiencies in cooperation between the IAF and IDF ground forces. But other reasons obtained as well. Because the IAF assigned a low priority to support of IDF ground forces before the Yom Kippur War, it did not develop much in the way of a C^2 infrastructure to ensure close cooperation with those forces.[41] And, for its part, IDF GHQ, convinced of the IAF's invincibility as a result of its performance in the Six-Day War, did not grasp the true limitations of airpower.

Combined Arms Warfare: The Naval Component

For the most part, the Israel Navy (IN) waged a solitary campaign, one that did not involve significant direct coordination with either IDF ground forces or the IAF. In the Mediterranean Sea arena, the navy's fast patrol boats (FPBs) armed with antiship missiles sank a number of their Syrian and Egyptian counterparts during the course of several battles, while incurring no losses of their own, driving the Syrian and Egyptian navies back into their ports and ensuring open sea lanes for maritime traffic into and out of Israel.[42] The IN's FPBs also bombarded shore targets along the Syrian and Egyptian coasts, thereby indirectly aiding IDF ground forces on both fronts, as the Syrian and Egyptian armies had to divert some forces to coastal defense. In the Red Sea arena, IN SOF conducted a series of destructive raids against the Egyptian navy, creating such turmoil that the Egyptians eventually abandoned their chief anchorage in the area.[43] This development was undoubtedly of some indirect assistance to IDF ground forces as they advanced southward toward the Red Sea during their cross-canal counterattack.

Summary

Critics of the IDF's practice of CAW during the Yom Kippur War tend to focus on how the war began rather than on how it ended. That the war cost the IDF over 100 aircraft, at least 700 tanks, up to 75 artillery tubes, and more than 2,500 soldiers, including a large number of officers, many of these losses incurred during the opening days of hostilities, is undisputed. Arab losses—mainly Syrian and Egyptian, but also Iraqi, Jordanian, and other—were much greater by the time hostilities ceased, however. Collectively, Arab forces lost 400–500 aircraft, 1,500–2,000 tanks, about 500 artillery tubes, and more than 15,000 soldiers.[44] Furthermore, territorially speaking, the IDF clearly had the upper hand at the end of the day.

Unquestionably, both the IDF's force structure and war-fighting doctrine before the Yom Kippur War were heavily slanted in favor of tanks and aircraft and against infantry, artillery, engineers, and logistics. And, undoubtedly, this nontraditonal force structure and war-fighting doctrine contributed significantly to serious IDF losses in tanks and aircraft during the first days of war. Still, these tank and aircraft losses were also incurred as a result of the IDF's general unpreparedness for war, which meant that it had to hurl its tanks and aircraft mainly unsupported at the Syrian and Egyptian armies in a desperate containment effort, because a few tank brigades and its fighter-bomber squadrons were the only forces it had immediately available to counter their offensives.

Whatever the case, to its credit, the IDF quickly improvised a force structure and war-fighting doctrine that allowed it to engage in traditional CAW at both the tactical and operational levels, at least insofar as concerned its ground forces. No less an authority than the officer in charge of the United State Army's Training and Doctrine (TRADOC) Command at the time of the Yom Kippur War remarked that:

> The Israelis demonstrated [that] it is possible to operate successfully in the face of highly lethal weapons by effective use of the combined arms team[:] that is[,] tanks supported by mechanized infantry, self-propelled artillery, and self-propelled air defense weapons.[45]

At the operational level, the IDF's counteroffensives into Syria and Egypt constituted impressive feats of traditional CAW.

Furthermore, even when traditional CAW was clearly absent in a direct sense, such as in the lack of close cooperation between the IAF and IDF ground forces throughout much of the war, it can still be argued that it existed in an indirect sense. The IAF, after all, kept Israel's skies clean, allowing the IDF's reserve formations to mobilize and deploy for battle swiftly. It also kept the Arab air forces pinned down throughout the war—through a combination of air-to-air combat and frequent air base attacks—so IDF ground forces were able to function essentially unimpeded by Arab air strikes. And, strategic attacks against targets in the Arab hinterland (e.g., against fuel depots and port facilities)—in which the IN also had a hand—served to divert at least some Arab forces away from the front, especially on the Golan.

Finally, it must also be noted that traditional CAW did not constitute a panacea for the IDF. This practice, to be sure, cut down on the IDF's losses in men and machines, but it certainly did not eliminate them. Even after it adopted a force structure and war-fighting doctrine compatible with traditional CAW, the IDF nevertheless still suffered substantial losses, particularly during the early stages of its counter-offensives into Syria and Egypt. In contrast to the Six-Day War, when the Syrian and Egyptian armies collapsed in the face of IDF pressure, they held their positions in the Yom Kippur War, often retreating in good order. This discipline, in tandem with the firepower at their disposal, meant that the IDF paid a high price for its victory in the war.

5

The Lebanon War

Combined Arms Warfare Reinvigorated

The Israel Defense Forces (IDF) won perhaps the greatest victory in its history in the 1973 Yom Kippur War. Caught by surprise on two fronts, its ground forces very thin along the front lines and its all-important reserve divisions not deployed for battle, facing adversaries whose combined forces were similar in scale to those at the disposal of the North Atlantic Treaty Organization (NATO) alliance in Europe at the time, the IDF was nevertheless able to contain the Egyptian and Syrian army offensives within days, to go over to the attack on both fronts shortly thereafter, and to defeat handily its adversaries within weeks. This swift and dramatic reversal of fortune has few equals in the annals of warfare.

Despite the IDF's triumph, however, the Yom Kippur War was a traumatic experience for the State of Israel. The dire situation both in the Sinai and on the Golan during the early days of the fighting, coupled with the heavy toll in men and matériel exacted by the hostilities, shook the country to its very core. Post-mortems of the fighting, therefore, typically focused on how the IDF fared at the start of the hostilities rather than on how it fared later in the war. Indeed, shortly after the end of the fighting, the Israeli government itself created an investigative body, known as the Agranat Commission, to look into why the IDF had failed to predict the outbreak of hostilities on the eve of the war, as well as why the IDF's forces had encountered such difficulty at the start of the war, particularly on the Sinai front.[1]

One positive side effect to emerge from the country's traumatic experience in the Yom Kippur War lay in the fact that it prompted the Israeli defense establishment to engage in some much-needed soul-searching, leading in turn to a thorough review of the IDF's force structure and war-fighting doctrine. In the 9 years between the Yom

Kippur War and the 1982 Lebanon War, while reaffirming its commitment to the central elements of the IDF's existing force structure and war-fighting doctrine, the defense establishment also introduced numerous reforms that were conducive to the practice of traditional (symmetrical) combined arms warfare (CAW) at the tactical and operational levels.

Combined Arms Warfare in the IDF Before the Lebanon War

Before the Yom Kippur War, the Israeli defense establishment subscribed to the idea that "the few" could overcome "the many" if the few were sufficiently superior to the many. After the Yom Kippur War, the defense establishment continued to endorse the notion that quality could overcome quantity, but it also decided to hedge its bet in this respect, drastically expanding the IDF's force structure over the following decade. Citing a few statistics that contrast the IDF's pre- and postwar force structures serves to illuminate the point.

IDF standing forces, composed of a mixture of professional soldiers and conscripts, grew by about 130 percent from the eve of the Yom Kippur War to the eve of the Lebanon War.[2] Reserve forces grew approximately 60 percent over the same period.[3] The quantity of tanks in the IDF's order of battle (ORBAT) doubled in the interim between the Yom Kippur War and the Lebanon War, while the quantity of modern armored personnel carriers (APCs) quadrupled during the same period.[4] The number of modern self-propelled artillery tubes grew by about 100 percent between the wars, while the number of fighter-bombers in the IAF's inventory increased by approximately the same percentage.[5]

This emphasis on quantity did not come at the expense of quality. The Israel Air Force (IAF) received the latest American fighter-bombers in the decade between the Yom Kippur War and the Lebanon War. The IAF also acquired highly sophisticated electronic countermeasures (ECM) gear, precision-guided munitions (PGMs), and unmanned aerial vehicles (UAVs), much of this equipment designed and produced by local industry. For their part, IDF ground forces took into inventory state-of-the-art tanks, APCs, artillery tubes, engineering vehicles, and infantry weapons, especially antitank missiles. And the Israel Navy (IN) continued to upgrade both the quantity and quality of its fast patrol boats (FPBs).

These quantitative and qualitative improvements were accompanied by an equally striking upgrade to the IDF's entire command and control (C^2) system. The Yom Kippur War C^2 system, whose performance was frequently poor, particularly on the Sinai front, was replaced by a far more sophisticated command, control, communications, and intelligence (C^3I) system that not only improved coordination among various ground forces headquarters (HQs), but also coordination among the ground forces, the air force, and the navy, the need of which had been one of the major lessons derived from the IDF's wartime experience.

Lest it be thought that the Israeli defense establishment concentrated primarily on modifying the IDF's force structure, it made similarly important changes to the IDF's war-fighting doctrine in the interim between the Yom Kippur War and the Lebanon War. On the one hand, in spite of the substantial armor and aircraft losses suffered by the IDF in 1973, the defense establishment remained committed to a war-fighting doctrine based on offensive maneuver warfare at the tactical and operational levels, with the tank and the fighter-bomber occupying the leading roles. The defense establishment simply did not buy into the idea that tanks and fighter-bombers had been rendered impotent by the appearance of advanced antitank and air defense weapons on the battlefield. The IDF chief of staff during the Yom Kippur War summed up this perspective on the issue at a symposium dedicated to the lessons of that war:

> The air force and armor remain the decisive factors on the battlefield. Their superiority and their success in battle are vital for victory in war. . . . [T]he decisive role of the air force and the armor is true in defense and offense.[6]

On the other hand, the Israeli defense establishment also digested the lesson that, on a battlefield saturated with advanced antitank and air defense weapons, tanks and fighter-bombers would function much more effectively and would absorb considerably fewer losses if they fought as part of traditional CAW task forces that incorporated ample artillery, infantry, engineering, and logistical support. Hence, the defense establishment's decision not only to beef up the IDF's artillery, infantry, and engineering arms, but also to train them specifically to work in unison with tanks and fighter-bombers in all-arms task forces. The aforementioned chief of staff remarked that:

> Combined operations and the combined task force present what I consider to be one of the major lessons to be learned on the tactical and [operational] levels. . . . [D]uring [the Yom Kippur

> War], we [i.e., the IDF] realized that combined operations have a growing importance in modern warfare, and are becoming a crucial factor for success in every single battle, in every campaign[,] and in every war. The reason for this is the development and variety of modern sophisticated weapons. For instance, the efficiency of anti-aircraft [sic] missile batteries has been proven during the [Yom Kippur] War. Because of the efficiency of the missiles, the task of our aircraft was complicated to a great extent, and many new difficulties and problems were introduced. It is therefore necessary to use different and combined weapons systems.
>
> Anti-tank [sic] missiles have also proved significantly effective in battle, although less so than the myth broadcast during the war. This does not in any way decrease the importance of tanks on the battlefield. On the other hand, there is no denying the fact that the missiles endanger tanks, and hinder their accomplishment of missions on targets defended by infantry. It is therefore mandatory to combine artillery support with every armor operation—this in order to neutralize anti-tank [sic] missiles and infantry as much as possible. There is also need for massive artillery support against the enemy's artillery, which represents a significant obstacle to the advance of armor.
>
> An armor operation must be accompanied by infantry in order to prevent anti-tank [sic] activity mainly in those areas where the tanks' visibility and fire are limited. To these should be added engineering crews, to cope with minefields, ditches[,] and other obstacles which, due to modern trends in the development of engineering equipment, such as quick mine-laying, have become quite numerous.[7]

The principles of traditional CAW, in other words, would guide IDF actions at the tactical and operational levels moving forward.

Air–ground cooperation constituted another important dimension of the IDF's war-fighting doctrine that displayed both continuity and change in the post-Yom Kippur War years. In respect of continuity, the IAF emphasized the attainment of air superiority over adversarial air forces and integrated air defense systems (IADSs) as its main objective in war; therefore, the bulk of its efforts in the interim between the Yom Kippur War and the Lebanon War went into upgrading its capabilities to shoot down fighter-bombers, attack air bases, and destroy air defenses, especially surface-to-air missile (SAM) batteries. Supporting IDF ground forces, particularly in the form of close air support (CAS) sorties, remained a distinctly secondary concern for the IAF. Perhaps surprisingly in light of their experience in the Yom Kippur War, the IDF's ground forces did not object to this ranking of priorities. Speaking on behalf of the ground forces, the aforementioned chief of staff observed that:

> The primary goal of the air force is to secure the skies throughout the country [i.e., over Israel] and above the combat forces. . . .
>
> I see the air force's main role in the support of ground forces in interdiction—to achieve destruction of the enemy's military infrastructure, cause havoc among troop movements[,] and, in one word, to paralyze the enemy forces. Even before 1973, I considered the subject of [CAS] the last priority task of the air force. I always believed that ground forces, secure from the enemy's air activity, should defeat enemy ground forces unaided. The [Yom Kippur War] reconfirmed my belief that [CAS] is costly in casualties, and that there is no positive ratio between relatively great losses and limited results.[8]

In respect of change, the IAF could not ignore the fact that, in a future round of hostilities, it might nevertheless be called upon to come to the assistance of IDF ground forces before it could achieve air superiority over an adversary's air force and/or IADS; therefore, it upgraded its capability to do so, principally by investing in the creation of a significant fleet of antitank helicopters.[9] Whether on offense or defense, these "flying tanks" were acquired in order to work hand in hand with IDF ground forces to defeat an adversary's armored and mechanized forces. Flying low and using the terrain to mask themselves, they would arrive quickly over the battlefield to knock out adversarial tanks and APCs with their antitank missiles, even in an environment where adversarial air defense weapons were present.

Finally, the Israeli defense establishment brought the IN into close coordination with IDF ground forces and the IAF in the aftermath of the Yom Kippur War. Prior to the war, the defense establishment had considered the IN to be essentially a coastal patrol force, whose main functions were to ensure that the sea lanes to Israel stayed open and to prevent a seaborne attack on the country; however, the navy's effective performance in the Yom Kippur War—driving the Egyptian and Syrian navies back into their ports and hammering strategic targets in those countries—convinced the defense establishment that the IN could work together with IDF ground forces and the IAF in a future war. Reflecting the defense establishment's sentiment in this regard, the aforementioned chief of staff commented that:

> . . . I changed my mind about the tasks of the [IN] after the [Yom Kippur War]. My belief [before the war] was that our navy, greatly outnumbered by the Arab navies, [had] only one task: to defend the coast. . . . Due to [its] excellent performance, I came to the conclusion that the [IN] can do much more in the event of another war.[10]

During the next round of hostilities, the defense establishment envisaged that the navy would directly assist IDF ground forces and the IAF to accomplish the country's wartime objectives.

Combined Arms Warfare in the Lebanon War

The IDF fought two distinct, albeit not entirely separate, campaigns during the Lebanon War: an asymmetrical campaign against Palestine Liberation Organization (PLO) forces, notwithstanding the fact that these forces had adopted many of the trappings of a standard national army, and a symmetrical (interstate) campaign against Syrian army forces. A rarity at this point in Israel's history, IDF forces boasted not only qualitative superiority vis-à-vis their adversaries, but also quantitative superiority. The IDF deployed approximately 75,000 soldiers, 1,250 tanks, 1,500 APCs, hundreds of artillery tubes, as well as hundreds of specialized armored vehicles attached to its engineering, maintenance, and logistical forces.[11] It arranged its ground forces into at least seven all-arms, division-sized task forces, with a majority of its manpower and armor opposite Syrian army forces in the Bekaa in southeastern Lebanon. These formidable ground forces were backed up by all of the IAF's and IN's assets.

Arrayed against IDF forces, the PLO deployed approximately 15,000 combatants, 300 tanks, 150 APCs, hundreds of artillery tubes and antitank missiles, as well as liberal quantities of antiaircraft artillery (AAA) and shoulder-fired SAMs. About half of the PLO's irregulars were stationed in southern Lebanon, with the other half located either in or around the Lebanese capital, Beirut. Syrian army forces in Lebanon consisted of approximately 25,000–30,000 soldiers, 300 tanks, 300 APCs, and hundreds of artillery tubes and antitank missiles, most of these forces stationed in the Bekaa. Once the war got underway, Syrian army general headquarters (GHQ) rushed additional men and equipment to Lebanon. An advanced IADS in the Bekaa that incorporated 19 SAM and scores of AAA batteries, as well as Syrian air force assets, protected these ground forces. The difficult natural and urban terrain in southern Lebanon, which both PLO and Syrian forces attempted to exploit to their benefit, offset to some extent the qualitative and quantitative superiority of IDF forces.

After several days of cross-border skirmishing, the IDF launched a full-scale offensive against PLO forces in southern Lebanon on June 6.[12] Four IDF task forces, one of them amphibious, crossed the border and swiftly advanced northward. Along the coast, an all-arms task force composed of tanks, mechanized and nonmechanized infantry, artillery, and engineers, acting according to the principles of traditional CAW, with infantry and artillery supporting armor and with engineers clearing obstacles on the roads and in urban areas, simply rolled over PLO forces in the area, quickly linking up with a task force composed of tanks and parachute infantry that had been landed behind PLO lines to the north. The story was much the same in the central and eastern sectors of southern Lebanon, where two more all-arms task forces composed of the same elements as the coastal task force, once more acting according to the principles of traditional CAW, easily crushed the PLO forces in their paths.

By the end of the first week of the war, all of the PLO forces in southern Lebanon had been either completely destroyed or thoroughly scattered, with a few minor exceptions, and IDF ground forces were at the gates of Beirut, where the PLO had located its GHQ. PLO losses in men and equipment in southern Lebanon were heavy, IDF losses in men and equipment light, in part because of the latter's efficient practice of traditional CAW at both the tactical and operational levels. It took IDF ground forces a few more days to subdue a number of isolated and bypassed pockets of resistance in urban areas of southern Lebanon, as they moved very slowly and deliberately in these areas in an effort to minimize collateral damage to civilian lives and property.

For the first few days of the war, the IDF largely steered clear of Syrian army forces in the central and eastern sectors of southern Lebanon.[13] Hostilities between IDF and Syrian forces were limited to clashes in places where Syrian forces were either in very close proximity to or actually intermingled with PLO forces. On June 9, however, the Israeli government and IDF GHQ decided to commence a full-scale offensive against Syrian forces in the Bekaa. Three IDF task forces composed of tanks, mechanized and nonmechanized infantry, artillery, and engineers launched a three-pronged offensive to smash Syrian defensive positions there. Adhering for the most part to the principles of traditional CAW at both the tactical and operational levels, these task forces mauled substantial chunks of two Syrian army tank divisions, as well as associated infantry and artillery forces, at limited cost to themselves during the next three days over terrain that heavily favored the defender. Cooperation between IDF tanks and antitank parachute infantry proved particularly effective in destroying Syrian armor, and engineers proved instrumental in overcoming natural and manmade obstacles in order to clear paths for tank and infantry forces to move forward.

The IDF offensive against the Syrian army, however, did not proceed entirely without fault. Israeli forces occasionally abandoned traditional CAW at the tactical level, engaging in either poorly or unsupported tank thrusts, leading in turn to setbacks and losses, albeit temporary and minor. Furthermore, at the operational level, the offensive unfolded rather slowly, largely as a result of confusion and disagreement within the Israeli government and IDF GHQ as to its scope and objectives, allowing considerable Syrian forces to escape the Israeli noose, as well as permitting the Syrian army to erect new defensive positions in eastern Lebanon. Nevertheless, the IDF had won a significant victory over its Syrian adversary by June 11, when the fighting came to a halt in the Bekaa.

The defeat of the Syrian army in this sector heralded the beginning of the long final phase of the Lebanon War, the Israeli siege of western Beirut, where the remnants of the PLO's forces, along with a Syrian army brigade, had embedded themselves among hundreds of thousands of Lebanese civilians.[14] From June 11 until June 25, IDF ground forces gradually tightened their siege on this part of the city by defeating PLO and Syrian army forces attempting to prevent them from doing so. IDF task forces composed of tanks, mechanized and nonmechanized infantry, artillery, and engineers once again employed traditional CAW to overcome their adversaries in a series of engagements in the vicinity of Beirut; however, neither the Israeli government nor IDF GHQ relished the prospect of entering the heart of the city itself to flush out PLO and Syrian forces, fearing that such a course of action would lead to massive troop losses, as well as to massive collateral damage to civilian lives and property.

The decision to initiate a siege of western Beirut rather than to storm it resulted in two months of heavy bombing and shelling of PLO and Syrian targets throughout western Beirut in an effort to pressure their forces into submission. The IAF, the IN, and IDF artillery forces all participated in this extended bombardment, but not in any sort of coordinated manner. The IDF, to put it another way, essentially abandoned traditional CAW during the siege. Predictably, the protracted bombardment of western Beirut caused heavy collateral damage, especially as PLO and Syrian forces cynically exploited Lebanese civil-

Figure 1.1 Throughout the first phase of the War of Independence, the practice of CAW among Israeli forces was essentially restricted to the integration of light weapons—mortars, machine guns, small arms, and IEDs—among infantrymen.

Figure 1.2 The IDF's practice of traditional CAW—integrating mechanized and nonmechanized infantry, tanks, artillery, and engineers—yielded positive results throughout the Sinai Campaign.

Figure 2.1 The IDF triumphs in the all-important break-in battles that determined the outcome of the Six-Day War on the Sinai front can be attributed largely to the effective practice of traditional CAW at the tactical level.

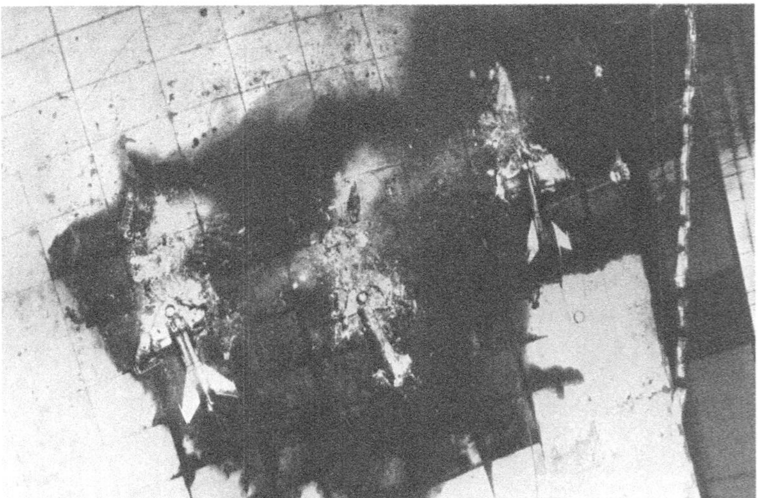

Figure 2.2 The IAF's destruction of the Egyptian, Jordanian, and Syrian air forces on the ground at the outset of the Six-Day War contributed significantly to the IDF's subsequent victories in the land battles on all three fronts.

Figure 3.1 For the most part, the IDF ignored traditional CAW during the War of Attrition, preferring instead to rely overwhelmingly on the IAF to prosecute the war against Egypt.

Figure 3.2 Though the IDF mostly ignored the practice of traditional CAW during the War of Attrition, many of its SPEC OPS constituted exceptions to the rule.

Figure 4.1 After the first few days of battle on the Sinai front during the Yom Kippur War, the IDF reverted to traditional CAW, integrating tanks, artillery, infantry, engineers, and fighter-bombers in respect of both its defensive and offensive operations.

Figure 4.2 The principles of traditional CAW were prominently in evidence during Operation Strongheart, the IDF's successful counter-crossing of the Suez Canal, which eventually saw large elements of three armored divisions fighting on the Egyptian side of the waterway.

Figure 5.1 For the most part, the IDF's initial offensives against PLO forces and the Syrian army during the Lebanon War adhered closely to the principles of traditional CAW.

Figure 5.2 During the protracted siege of Beirut, the IDF essentially abandoned the practice of traditional CAW, instead relying on air strikes and artillery bombardments to pound its adversaries into submission.

Figure 6.1 IDF armored forces suffered significant losses at the hands of Hizbullah's antitank missiles during the Second Lebanon War, because the IDF failed to support its tanks adequately with infantry, artillery, and engineering forces.

Figure 6.2 The IDF's renewed commitment to traditional CAW at the tactical and operational levels in the aftermath of its lackluster performance in this regard during the Second Lebanon War paid substantial dividends during Operations Cast Lead and Protective Edge.

Figure 7.1 Since the Six-Day War, air–ground cooperation has played a major role in IDF operations during LICs, with helicopters frequently being employed to transport infantrymen to their targets.

Figure 7.2 The IDF's infantry has occupied center stage in Israel's efforts to counter LICs in the twenty-first century.

Figure 8.1 The smooth cooperation between the IAF and IDF SOF was one of the major reasons behind the success of Operation Jonathan.

Figure 8.2 The IAF destroyed the Syrian nuclear weapons production facility at al-Kibar, as these before and after photos of the installation clearly show, with the assistance of IDF cyber warfare assets.

ians as human shields. Equally predictably, the siege resulted in international outrage being directed at Israel, as well as a loss of support for the war among the Israeli populace. Yet, in the end, the protracted bombardment of western Beirut, in tandem with international diplomacy, paid dividends, as both PLO and Syrian forces agreed in mid-August to abandon their positions in the city.

While the siege of western Beirut was undoubtedly a messy and destructive affair, it should not obscure the fact that the IDF chalked up substantial achievements during the Lebanon War. In addition to losing as many as 2,000 combatants (even more if allied Lebanese militiamen and foreign terrorists are counted among the total), PLO forces lost all of their heavy equipment (all of their tanks, all of their artillery tubes, all of their antiaircraft guns, and so on), which was either destroyed or captured by the IDF.[15] Moreover, PLO forces were ejected from Lebanon, never to return there. The Syrian army lost at least 1,000 soldiers (probably more) and 300–350 tanks. For its part, the IDF lost approximately 370 troops and 30–40 tanks, the majority of these losses in men and machines incurred during fighting with the Syrian army.

Combined Arms Warfare: Air–Ground Cooperation

The IAF executed one of the most spectacular large-scale operations in its history during the Lebanon War. On June 9, it carried out Operation Mole Cricket 19, which resulted in the swift destruction of the Syrian IADS in the Bekaa in a brilliant example of traditional CAW at the tactical and operational levels.[16] The IAF apparently inflicted much of the critical damage to the Syrian air defense system in the first 10–20 minutes of the operation, then mopped up the remaining components during several hours of follow-on strikes. At the tactical level, IAF fighter-bombers fitted with PGMs and ECM pods, EW aircraft, intelligence-gathering UAVs, and decoy drones, all under the control of a highly sophisticated C^3I system, worked seamlessly together to eliminate the IADS' C^2 system in short order. Successive strikes by fighter-bombers fitted out with PGMs, general purpose (GP) bombs, and cluster bomb units (CBUs) then methodically destroyed SAM and AAA batteries alike. At the operational level, the IAF coordinated its attack with IDF long-range artillery forces. The latter, utilizing both long-range guns and surface-to-surface missiles (SSMs), made a substantial contribution to the maelstrom unleashed on the Syrian IADS. Once the smoke cleared over the Bekaa battlefield, it quickly became evident that at least 17 SAM batteries, as well as scores of AAA batteries, had been reduced to smoking ruins as a result of the IDF's air–artillery onslaught. The fact that the IAF did not lose a single aircraft in Operation Mole Cricket 19 rendered the accomplishment even more spectacular. The IAF would go on to destroy 12 additional Syrian SAM batteries in the Bekaa and, possibly, inside Syria itself (the latter through standoff attacks launched from outside of Syrian airspace) later in the war.[17]

Syrian aircraft that tried to disrupt the operation were engaged by IAF aircraft, which shot down more than 20 Syrian machines without loss to themselves on June 9 alone.[18] Indeed, throughout the war, the IAF shot down 90–100 Syrian fighter-bombers and helicopters in air-to-air combat without losing a single aircraft of its own.[19] The IAF's losses in the war amounted to two fighter-bombers and three helicopters, all brought down by antiaircraft fire.[20]

The IAF attained air superiority everywhere in Lebanese airspace as a result of the devastating outcome of Operation Mole Cricket 19 in combination with its dominance in air-to-air combat. Consequently, IDF ground forces never came under sustained air attack. The sporadic close air support (CAS) and interdiction strikes mounted by the Syrian air force inflicted only minor damage on IDF ground forces throughout the fighting. IAF air superiority not only assisted the IDF ground forces in this indirect sense at the operational level, but also in both a direct and indirect sense at the tactical level.

Once Syrian air defenses and aircraft had been removed from the battlefield, the IAF was free to unleash its own interdiction and CAS campaigns on behalf of IDF ground forces. Interdiction strikes were primarily directed against transportation infrastructure, namely roads and bridges.[21] These strikes inhibited the movement of PLO and Syrian manpower and equipment around Lebanon, thereby assisting IDF ground forces in an indirect sense. IAF interdiction strikes also occasionally mauled PLO and Syrian forces in transit, thereby assisting IDF ground forces in a direct sense. On June 10, to cite one notable instance, IAF fighter-bombers caught a Syrian armored brigade in the open before it could reach the battlefield, destroying at least 20 of its tanks and scattering the rest, which severely degraded its capability to contribute on the Bekaa battlefield.[22]

The results of the IAF's CAS campaign are not as clear-cut.[23] On the one hand, the IAF frequently could not provide CAS to advancing ground forces in urban areas (because of concerns about collateral damage) and in mountainous areas (because of terrain obstacles). On the other hand, however, when the IAF could provide CAS to the ground forces, it generally did so rather effectively, much more effectively than in the Yom Kippur War. UAVs that furnished intelligence information in real time, a highly sophisticated C^3I system that included air–ground links, and advanced aerial platforms and munitions combined to ensure accurate and timely CAS sorties against PLO and Syrian forces in many instances, though mishaps, including several friendly fire incidents, were not unknown. The IAF's attack helicopter fleet appeared to be especially adept at picking off tanks and other armored vehicles that posed a threat to advancing ground forces, particularly in difficult terrain.

Finally, the IAF assisted IDF ground forces logistically. The air force employed both transport aircraft and helicopters to ferry soldiers and equipment to the front, as well as to ferry wounded back to Israel. And it employed helicopters to insert parachute infantry behind the lines on a number of occasions during the early days of the war.[24]

Combined Arms Warfare: The Naval Component

The IN participated in traditional CAW at both the tactical and operational levels during the Lebanon War.[25] At the tactical level, IN warships furnished direct fire support to IDF ground forces moving up the Lebanese coast, helping to reduce PLO strongpoints along the way. At the operational level, IN transport vessels landed a task force composed of tanks and parachute infantry behind PLO lines early in the war, and they subsequently reinforced this beachhead by bringing in additional men and equipment shortly thereafter. Furthermore, the IN's blockade of the entire Lebanese coastline prevented the PLO from bringing men and equipment into the country via the sea, thereby assisting IDF ground forces in an indirect sense.

Summary

In the aftermath of the Yom Kippur War, the Israeli defense establishment revised the IDF's force structure and war-fighting doctrine to align better with the requirements of traditional CAW. In the interim between the Yom Kippur War and the Lebanon War, the defense establishment not only expanded and improved previously neglected arms, such as infantry and artillery, but also carefully integrated them into the IDF's war-fighting doctrine at both the tactical and operational levels.

The fruits of these reforms became evident during the Lebanon War. Though the IDF's performance in the war was certainly not flawless at either the tactical or operational level, Israeli ground, air, and naval forces nevertheless registered major accomplishments in the fighting. The IDF thrashed both PLO and Syrian forces in southern Lebanon in urban and mountainous terrain that strongly favored the defender. Much of the IDF's success in battle is directly attributable to the embrace of traditional CAW at both the tactical and operational levels. Mutual support among different arms was the dominant trait of the battles waged by the ground forces. Cooperation between the IAF and the IN, on one the hand, and IDF ground forces, on the other hand, was also the order of the day throughout the hostilities.

The Lebanon War was the last interstate war (IW) fought by the IDF to the present day. In the decades since that war, Israel's strategic environment has undergone fundamental change. Once hostile neighboring Arab states have either made their peace with Israel (Egypt and Jordan) or disintegrated into utter chaos to the point where they essentially exist in name only (Syria and Iraq). In the twenty-first century, therefore, IW has not constituted the paramount threat to Israeli national security. Instead, the IDF has had to come to grips with the phenomenon of asymmetrical war (AW), in which its adversaries have been nonstate actors. This shift in the nature of conflict, in turn, has had considerable ramifications for both the IDF's force structure and war-fighting doctrine.

6

The Second Lebanon and Gaza Wars

Combined Arms Warfare and Asymmetrical War

Since the middle of the twenty-first century's opening decade, the State of Israel has fought three wars: the 2006 Second Lebanon War (or Operation Change of Direction), the 2008–9 Gaza War (or Operation Cast Lead), and the 2014 Gaza War (or Operation Protective Edge). These wars differed from the country's previous wars in that its adversaries in these conflagrations were exclusively nonstate actors. Though the Israel Defense Forces (IDF) had faced nonstate actors in the 1947–49 War of Independence and the 1982 Lebanon War, both of these conflagrations were dominated by their interstate rather than asymmetrical character. Israel in the twenty-first century, in short, had entered into a new era in respect of its national security experience, an era of asymmetrical war (AW).

Twenty-four years separated Israel's last interstate war (IW) from its first AW. During that quarter century, the threat to Israel posed by surrounding Arab states had declined rather dramatically. Egypt and Jordan had signed peace treaties with Israel in 1979 and 1994, respectively, while Iraq had essentially been eliminated from the anti-Israel axis in consequence of the 1991 Gulf War and the 2003 Iraq War. Concomitantly, nonstate actors such as Hizbullah, Fatah, Hamas, and Islamic Jihad had replaced the Arab states as the most immediate threat to Israel. Iran had become a very dangerous thorn in the country's side, too, both directly—as a consequence of its expanding ballistic missile and nuclear weapons programs—and indirectly—as a consequence of its generous support of the abovementioned nonstate actors, particularly Hizbullah. Moreover, the IDF had witnessed the outcome of the Gulf War between the American-led coalition and the Iraqi army, as well as the outcome of the 1999 Kosovo War between North Atlantic Treaty Organization (NATO) forces and the Serbian army, in which high-technology, standoff firepower had been the most pronounced trait in the coalition and NATO victories. All of these developments in Israel's strategic environment coalesced between the Lebanon War and the Second Lebanon War, prompting a major revision to the IDF's "way of war" during those years.

Combined Arms Warfare in the IDF Before the Second Lebanon War

The same three variables that had largely convinced the Israeli defense establishment to adopt offensive maneuver warfare in the early 1950s gradually drove it to adopt offensive attrition warfare (with an enhanced emphasis on defense as well) in the interim between the two Lebanon wars.[1] Swift and deep thrusts into an adversary's hinterland by mobile columns had become passé. What worked to defeat national armies would not necessarily produce victory over nonstate military forces. Moreover, as a result of its experiences (in Lebanon as well as Judea, Samaria, and Gaza), Israel no longer had an interest in seizing and controlling additional territory inhabited by hostile populations. The application of massive, precision firepower, especially from the air, reasoned the Israeli defense establishment, seemed to provide a much more efficient way to cope with nonstate actors like Hizbullah and Hamas. Massive, precision firepower from the air could also be used to address the Iranian threat. Numbers of men and machines, now heavily in the IDF's favor, also augured well for offensive attrition warfare, as did the fact the no foreign powers, including the Arab states, would—or even could—come to the immediate assistance of nonstate actors like Hizbullah or Hamas. That these actors would essentially be on their own against Israel, reckoned the Israeli defense establishment, would give the IDF more time to accomplish its objectives than it had during IWs. The IDF could now use its superior numbers to whittle down its adversaries in a comparatively leisurely manner.

The shift from an emphasis on offensive maneuver warfare to an emphasis on offensive attrition warfare significantly affected the IDF's force structure and war-fighting doctrine. As the threat of IW receded further and further into the background, the Israeli defense establishment concluded that the IDF should become "smarter and slimmer." To this end, it slowly but unmistakably reduced its reliance on heavy ground forces—that is, tanks and mechanized infantry. The IDF steadily cut the number of tank and mechanized infantry brigades at its disposal, a trend that has continued to the present day.[2]

Simultaneously, the IDF steadily built up those of its light ground forces better attuned to engage nonstate actors on the battlefield; hence, elite infantry forces, special operations forces (SOF), engineering forces, and intelligence forces were enhanced at the expense of tank and mechanized infantry forces.[3] In the post-Lebanon War decades, the Israeli defense establishment particularly favored the Israel Air Force (IAF) in its force structure plans, another trend that has continued to the present day.[4] The IAF received not only the most advanced fighter-bombers in the world, but also the most advanced electronic warfare (EW) systems, the most advanced precision-guided munitions (PGMs), and the most advanced unmanned aerial vehicles (UAVs), all of them tied together by a state-of-the-art battle management system.

These far-reaching changes to the IDF's force structure were accompanied by equally far-reaching changes to its war-fighting doctrine. Most significantly, whereas firepower had generally taken a back seat to mobility in the IDF's war-fighting doctrine before the Lebanon War, mobility generally took a back seat to firepower in the interim between the two Lebanon wars, yet another trend that has continued to the present day.[5] Under the influence of the Revolution in Military Affairs (RMA), as exemplified in the Gulf and Kosovo wars, the Israeli defense establishment became enamored of long-range, precision firepower. The defense establishment believed that, not only could such firepower tackle effectively distant threats, like the one posed by Iran, whose forces could launch ballistic missiles at Israel, but also that it could tackle effectively local threats, like the ones posed by Hizbullah and Hamas, whose forces often used urban terrain as cover for their rocket and other attacks. Ground forces, of course, were also necessary to counter such nonstate actors; however, heavy ground forces, which were optimized for IW, now took a back seat to light ground forces, which were optimized for low-intensity conflict (LIC). Tanks and mechanized infantry still had roles to play in an era in which Israel faced insurgent/terrorist groups rather than national armies; however, they were now relegated to secondary roles in support of elite infantry forces, SOF, and engineering forces, all of which were able to operate more nimbly in the urban and mountainous terrain in which the IDF found itself face-to-face with the insurgent/terrorist groups.

The changes to its force structure and war-fighting doctrine in the interim between the two Lebanon wars had significant consequences in respect of the IDF's capability to adhere to the principles of traditional combined arms warfare (CAW). The LICs in which the

IDF found itself during these decades were fought mainly by infantry forces, supplemented when necessary by precision fire from the air and ground; therefore, as it steadily shifted its focus to counter nonstate actors, the IDF eventually forfeited to a great extent (through atrophy) its capability to wage traditional CAW at the both the tactical and operational levels, with the deterioration in the ability of its heavy ground forces to engage in this type of fighting becoming particularly evident over the years. Air–ground cooperation at the both the tactical and operational levels also largely went out of the window. By the outbreak of the Second Lebanon War, then, the IDF as a whole was in no shape, in terms of force structure and war-fighting doctrine, to wage traditional CAW at the tactical and operational levels.

Combined Arms Warfare in the Second Lebanon War

The Israeli government decided to withdraw the IDF unilaterally from southern Lebanon in the spring of 2000, reasoning that the army's 18-year stay in the "Lebanese mud" had outlived its usefulness to the country's national security interests. The government expected the IDF's withdrawal to result in calm along Israel's northern border, as Hizbullah's raison d'être, the "liberation" of Lebanese soil from the "Zionist occupier," had now been realized. Hizbullah, however, simply changed its raison d'être to the "liberation" of Palestine—that is, the destruction of Israel—in order to justify its continued existence and belligerence.

From 2000–6, therefore, Hizbullah continually provoked Israel with cross border attacks, ostensibly in support of Palestinian "resistance" during the 2000–5 Second Intifada. Caught up in a LIC with Hamas, Islamic Jihad, Fatah and other Palestinian insurgent/terrorist groups, successive Israeli governments preferred not to escalate hostilities on the northern border, opting instead for a tit-for-tat policy. That is, until a particularly deadly Hizbullah raid on July 12, 2006 altered the rules of the game.

Later that same day, the Israeli government authorized the IDF to undertake a massive assault against Hizbullah, thereby dragging the country into an AW for which it had not properly prepared itself.[6] For starters, the Israeli government had not previously thought out what objectives the country could realistically obtain during such a war, let alone how to obtain them. And, for its part, the Israeli defense establishment had not developed beforehand a coherent and integrated war plan for waging an AW. Consequently, the Israeli war effort lacked both focus and steam, quickly devolving into two essentially separate campaigns: one in the air, the other on the ground.

The primary threat to Israel posed by Hizbullah stemmed from the latter's possession of a substantial arsenal of long-, medium-, and short-range rockets, which could strike cities, towns, and villages throughout the north and center of the country. The air campaign was intended in large part to eliminate this threat, at least insofar as concerned the long- and medium-range rocket arrays.[7] It did so in grand style. Hizbullah's long-range rocket array was knocked completely out of action early in the war, before any of these rockets could be fired into Israel.[8] And its medium-range rocket array was crippled in the fighting, while inflicting only limited damage inside Israel.[9] Almost all of Hizbullah's fixed medium- and long-range rocket launchers were destroyed by IAF strikes during the nights of July 13 and 18, respectively, in highly effective displays of traditional CAW at the tactical level that saw not only various air force assets (fighter-bombers, UAVs, reconnaissance satellites, and so on), but also intelligence and, most likely, SOF assets work seamlessly together to accomplish their missions. Furthermore, IAF counterbattery fire during the fighting was so rapid and devastating that Hizbullah's mobile medium-range rocket batteries were often destroyed after firing only a single rocket from their multi-barrel launchers.

The other facets of the Israeli air campaign bore much less fruit, mainly as a result of a lack of air–ground cooperation at both the tactical and operational levels throughout the war. The low signature of Hizbullah's short-range rocket array—these rockets could be easily moved and hidden—made it next to impossible to destroy them in great numbers from the air alone. Dedicated and sustained cooperation between air and ground forces to ferret out and engage the rocket launchers was required to accomplish this task; however, no such cooperation occurred during the war. Hizbullah, therefore, was able to launch thousands of short-range rockets into northern Israel throughout the fighting, causing considerable damage there. Similarly, repeated IAF strikes against Hizbullah's headquarters (HQ) facilities in Beirut, as well as against Lebanon's transportation infrastructure, had no effect on the group's ability to prosecute the war, in part because IDF ground forces never exploited to their advantage the heavy damage done to these targets.

The air campaign, much of which had been planned in advance, unfolded rather smoothly in comparison to the ground campaign.[10] Consistent with its embrace of offensive attrition warfare based primarily on aerial firepower, the Israeli defense establishment was not keen to unleash the IDF in a major "ground maneuver" in the Second Lebanon War. Indeed, the defense establishment had not even prepared a contingency plan for a large-scale incursion prior to the war. For most of the Second Lebanon War, therefore, the IDF ground campaign consisted primarily of shallow raids into southern Lebanon in order to destroy Hizbullah's military infrastructure in the area and to engage its combatants in firefights. Frustrated with the slow progress of the ground campaign during the first weeks of fighting, the IDF gradually expanded the scope and depth of these raids in subsequent weeks, though its forces still did not stray very far from the Israeli border.

Generally speaking, the IDF got the better of Hizbullah during these raids. Nevertheless, problems in respect of traditional CAW quickly became evident.[11] Coordination among different arms—tanks, infantry, artillery, and engineers—was often deficient, leading to losses in men and machines. Logistically speaking, the IDF experienced a lot of trouble keeping its ground forces in the field supplied with the necessary consumables, so much so that the IAF was called upon to shoulder much of this burden through airdrops. And, while the IAF engaged in an often effective close air support (CAS) effort on behalf of the ground forces with its attack helicopter fleet, neither IDF ground forces nor the air force itself was satisfied with the quality of air–ground coordination demonstrated in the war.

The real extent of the IDF's regression in respect of the practice of traditional CAW at the tactical and operational levels, however, would only reveal itself during the final days of the war, when the Israeli government authorized a large-scale ground offensive to seize all of southern Lebanon up to the Litani River. In spite of the fact that it inflicted substantial casualties on Hizbullah forces, this brief, slow-moving, and, occasionally, bumbling offensive bore no resemblance to its predecessors in the 1967 Six-Day War, the 1973 Yom Kippur War, and the 1982 Lebanon War. This offensive, in contrast to earlier ones, was not characterized by swift thrusts of all-arms task forces deep into the adversary's rear areas under an umbrella of airpower. Indeed, the most significant contribution made by the IAF to the offensive lay in its airlifting of troops and consumables to the front line. On the ground, cooperation among tanks, infantry, artillery, and engineers was at least as poor—probably even poorer—than it had been earlier in the fighting. In one especially egregious incident, a tank column without proper infantry support attempted to advance in hilly terrain only to be shot to pieces in a Hizbullah antitank ambush that claimed approximately half of the force.[12]

Despite the IDF's undeniable shortcomings in regard to the practice of traditional CAW, it nevertheless carried the day in the Second Lebanon War. The IDF certainly did not achieve a swift, elegant, and decisive victory à la its Six-Day War victory. While Hizbullah had certainly been bloodied by the time a cease-fire took effect on August 14, it had not been beaten to a pulp. Rather, the IDF had won the Second Lebanon War on points, not by a knockout. Hizbullah's arsenal of long- and medium-range rockets had been eviscerated, as had the lion's share of its military infrastructure in southern Lebanon. And the group had lost at least five times as many combatants in the fighting as had the IDF, many of them in close-range firefights. The fact that the Israel–Lebanon border has been quieter since the end of the Second Lebanon War than at any other time since the end of the Six-Day War speaks eloquently to the IDF's accomplishments in the war.

Combined Arms Warfare in the IDF Before Operation Cast Lead

These accomplishments notwithstanding, the Israeli defense establishment was badly shaken by the IDF's performance in the Second Lebanon War, triggering a thorough review of its force structure and war-fighting doctrine.[13] Commissions both inside and outside of the IDF were set up to assess its shortcomings in the war and to recommend fixes to those shortcomings. These commissions essentially reaffirmed the pre-existing emphasis on offensive attrition warfare in an era characterized by AW and LIC; however, they also strongly advocated on behalf of reforms to the IDF's practices in respect of ground maneuvers and traditional CAW. The Israeli defense establishment, in other words, conceded that the IDF had placed too much faith in firepower and too little faith in mobility to accomplish battlefield objectives, as well as too little stress on maintaining the skills necessary to wage traditional CAW at the tactical and operational levels.

The IDF's force structure remained largely unaffected by the defense establishment's review and recommendations. Procurement trends in evidence prior to the Second Lebanon War continued to dominate the IDF's arms acquisitions moving forward—that is, the IAF and light ground forces continued to be favored at the expense of heavy ground forces. Consequently, the number of tank and mechanized infantry brigades in the IDF's order of battle (ORBAT) continued to shrink, though the quality of the remaining forces improved as a result of more intensive and more realistic training, particularly in the art of traditional CAW.

The IDF's war-fighting doctrine, on the other hand, underwent considerable revision. Consistent with its belief that the IDF had relied too heavily on firepower, especially aerial firepower, at the expense of mobility in the Second Lebanon War, the Israeli defense establishment moved the needle on the firepower–mobility spectrum. A substantial ground maneuver, vowed the defense establishment, would complement an air campaign in any future AW. It must be said, however, that the enhanced role of the ground maneuver in the IDF's war-fighting doctrine did not signal a return to the days of offensive maneuver warfare. A future ground maneuver was envisaged to be a shallow, slow-moving affair designed to grind down the adversary, not to penetrate rapidly and deeply into its rear in order to bring about its collapse. Hand in hand with the enhanced role of the ground maneuver in the IDF's war-fighting doctrine went a renewed emphasis on the practice of traditional CAW, not only within the ground forces, but also among air, naval, and ground forces. This recommitment was further reinforced by the IDF's decision to jettison all of the many vacuous concepts and terms, many of them influenced by the obscurantism characteristic of postmodernist philosophy, which had crept into its lexicon in the early years of the twenty-first century, in favor of the traditional and meaningful concepts and terms associated with twentieth-century warfare. Doctrinally speaking, in short, the IDF went "back to basics" in the years between the Second Lebanon War and Operation Cast Lead.

Combined Arms Warfare in Operation Cast Lead

Rocket fire constituted the primary threat to Israel from Gaza before Operation Cast Lead, just as it had from Lebanon before the Second Lebanon War. Ever since the unilateral Israeli withdrawal from Gaza in 2005, the quantity and quality of rockets fired at southern Israeli cities, towns, and villages had grown steadily. On the eve of Operation Cast Lead, the rocket arsenal of the Hamas-led coalition of insurgent/terrorist groups in Gaza, while not nearly as formidable as the one possessed by Hizbullah two years earlier, had the potential to inflict serious harm on the Israeli populace in the south of the country.

The Israeli defense establishment recognized that it was only a matter of time before it would be necessary for the IDF to fight a war in Gaza in order to curb the rocket fire. So as not to be caught once more unready to respond to the challenge of AW, therefore, the defense establishment engaged in meticulous preparations for Operation Cast Lead well in advance of its actual commencement. Politically speaking, the Israeli government formulated clear and realistic objectives to be obtained in a war with nonstate actors in Gaza, namely the crippling of their military capabilities, particularly those related to rocket fire, and the reinforcement of Israeli deterrence. Entirely absent was the sort of loose, spur-of-the-moment talk about eviscerating once and for all such insurgent/terrorist groups, the sort of talk that had made it seem as if Israel had not chalked up any tangible accomplishments as a result of the Second Lebanon War. Militarily speaking, the IDF drew up a coherent and integrated war plan far in advance of its implementation. Consistent with its renewed foci on ground maneuver and traditional CAW, the plan called for an initial high-intensity air campaign against high-value targets to be supplemented shortly thereafter by a major ground campaign to finish the job.

By the final days of 2008, conditions had become ripe for the IDF to launch Operation Cast Lead. The previous few weeks had witnessed a considerable escalation of Hamas-led rocket fire into southern Israel such that normal everyday life in the cities, towns, and villages there had been disrupted to an unacceptable extent. Hence, on December 27, the Israeli government authorized the IDF to commence the operation.[14] The first phase of the campaign—the aerial phase—would last approximately one week. A massive opening blow on the first day of the war killed or wounded hundreds of Hamas insurgents/terrorists and completely smashed the Hamas-led coalition's command and control (C^2) infrastructure, which essentially ceased to exist for the remainder of hostilities.[15] The IAF quickly broadened the scope of the air campaign to include rocket launchers, arms depots, rocket- and mortar-manufacturing facilities, training bases, leadership hideouts, arms smuggling tunnels, and various other targets associated with the coalition's military capabilities. The week-long Israeli air campaign, in short, severely degraded the Hamas-led coalition's military capabilities. The close cooperation among IAF assets, military (and civilian) intelligence assets, and SOF assets at the tactical and operational levels lay behind this impressive feat of traditional CAW.

On January 3, the IDF launched the second phase of Operation Cast Lead, a two-week combined air–ground campaign against the coalition.[16] Light ground forces—that is, elite infantry forces, SOF, and engineering forces—took point, with heavy ground forces—that is, tanks and artillery—in support. The IDF's ground maneuver was deliberately slow and shallow in a painstaking attempt to grind down further the Hamas-led coalition's military capabilities. To that end, the same broad range of targets attacked during the first phase of Operation Cast Lead was also attacked during the second phase of the operation. The IDF placed a special emphasis during this phase of the operation on pushing back the coalition's rocket-launching sites from the immediate proximity of the Israel–Gaza border in order to reduce the rockets' coverage of Israeli territory.

Throughout the joint phase of the operation, the IDF adhered strictly to the principles of traditional CAW at both the tactical and operational levels. Heavy ground forces ably assisted light ground forces at the tactical level, the former providing effective fire support before and during the latter's assaults on coalition strongpoints. IAF air controllers were fully integrated into the ground forces and IAF assets—that is, attack helicopters and UAVs—were at the direct disposal of ground commanders upon demand.[17] One IAF officer remarked that the "concentration of air assets . . . permitted unparalleled air–land cooperation."[18] In consequence, the IAF's CAS effort in Operation Cast Lead was quite effective, routinely knocking out coalition strongpoints and detonating improvised explosive devices (IEDs) just in front of advancing IDF ground forces. At the operational level, continuous IAF strikes against arms smuggling tunnels, as well as vigorous Israel Navy (IN) patrols of the coastal waters adjacent to Gaza, ensured that the Hamas-led coalition could not bring in additional men and machines from abroad with which to confront IDF ground forces.

Operation Cast Lead came to an end on January 18, a unilateral Israeli cease-fire followed quickly by a unilateral coalition cease-fire. In contrast to the outcome of the Second Lebanon War, the outcome of Operation Cast Lead was never in doubt. The IDF had scored a clear-cut victory in the fighting. Rocket fire into Israel had slackened from an average of 60 strikes per day at the start of hostilities to an average of 20 strikes per day at the close of hostilities.[19] Moreover, the extent of territory in southern Israel under fire dwindled over the course of the operation as the rocket-launching sites were pushed back from the border. All in all, coalition rocket fire into southern Israel during Operation Cast Lead caused very few deaths and very little damage. Similarly, IDF losses in men and machines were minuscule in comparison to those suffered by the Hamas-led coalition, most of whose military capabilities had been smashed to pieces. Not only had Israel met its first objective—inflicting a devastating blow to the coalition's military capabilities—but also it had met its second objective—bolstering Israeli deterrence along the Israel–Gaza border.

Combined Arms Warfare in the IDF Between the Gaza Wars

The Israeli defense establishment believed that the first round of AW in Gaza would not be the last one. Only the timing of a second round of war between the IDF and the Hamas-led coalition, to its way of thinking, remained at issue. Because its approach to AW had borne fruit during Operation Cast Lead, the defense establishment decided to stick with it in the years following the operation. Indeed, according to an Israeli defense analyst:

> [T]he IDF's operational concept underwent no fundamental change [after] Operation Cast Lead. It was based on [the use of firepower], including precise standoff firepower[,] against previously selected targets and the intensification of damage to incidental targets. In addition to [firepower], the IDF maintained the readiness of ground forces to undertake a limited ground incursion with the aim of destroying the military infrastructure of Hamas and other armed groups operating in [Gaza].[20]

During the next round of AW, in other words, the IDF would once again open with an "air-only" campaign followed by an air–ground campaign in the broader context of offensive attrition warfare. One new wrinkle fully consistent with this approach to AW lay in the addition of an antirocket interceptor system to the IDF's arsenal, a requirement for which had been one of the lessons the defense establishment had drawn from the Second Lebanon War.

Combined Arms Warfare in Operation Protective Edge

The potency of deterrence has a tendency to diminish over time when the actor being deterred is an insurgent/terrorist group fanatically committed to its ideology. As the healthy dose of realism administered during a previous battlefield defeat steadily wares off over time, it is replaced once more by the growing fever of ideology. And so it was in the case of the Hamas-led coalition. Unwilling to countenance the very existence of a Jewish state, let alone peaceful coexistence with it, the Hamas-led coalition eventually returned to the battlefield. In the years following Operation Cast Lead, cross-border flare-ups became more frequent and more intense over time as the operation receded into history. A particularly notable flare-up in late 2012—an extremely intense week-long exchange of firepower between the IAF and the coalition dubbed Operation Pillar of Defense by Israel—that ended in a drubbing for the latter temporarily halted the slide toward another round of war by buttressing Israeli deterrence; nevertheless, by the summer of 2014 conditions had become ripe for a second round of AW between the IDF and the Hamas-led coalition. A single match would be sufficient to set off a major conflagration.

That match appeared in the form of a deadly incident in the Judea–Samaria region in which Hamas operatives kidnapped and murdered a number of Israeli youths. In response, Israel embarked on a large-scale crackdown on the group's operatives in the region, which in turn triggered renewed rocket fire from Gaza. In an attempt to avoid war, the Israeli government initially exercised restraint in answering the rocket fire, hoping that the Hamas-led coalition would reciprocate Israel's gesture; however, rather than respond in kind, the coalition continued to escalate its rocket attacks. Hence, on July 8, the Israeli government authorized the IDF to commence Operation Protective Edge.[21]

In the years following Operation Cast Lead, the Hamas-led coalition significantly improved its military capabilities.[22] First, its rocket arsenal expanded both quantitatively and qualitatively speaking, incorporating munitions that could reach all the way to Tel Aviv and Jerusalem. Second, the coalition constructed a network of over 30 "attack" tunnels, approximately half of them penetrating into Israeli territory, which insurgents/terrorists could use to launch surprise assaults. And, third, the coalition's forces were better trained and better equipped than before to wage a defensive campaign on the ground inside Gaza. The IDF, to put it another way, would face a much tougher test in its second round of AW in Gaza.

The first phase of Operation Protective Edge, as called for in the IDF's approach to AW, opened with an air campaign that struck the same types of target hit in Operation Cast Lead's air campaign: C^2 infrastructure, rocket launchers, arms depots, rocket- and mortar-manufacturing facilities, training bases, leadership hideouts, attack and arms smuggling tunnels, and so on. Thousands of targets were destroyed or damaged in this high-intensity campaign characterized by close cooperation at the tactical and operational levels among IAF assets, military (and civilian) intelligence assets, and SOF assets. Traditional CAW was once more the order of the day.

The second phase of Operation Protective Edge, which began on July 18 and ended on August 3, saw IDF ground forces launch a shallow and slow-moving incursion into Gaza. This time around, the ground maneuver was less about countering the rocket threat, as the

IAF's antirocket interceptor system was proving to be highly effective in protecting inhabited areas from rocket strikes, than about knocking out the coalition's network of attack tunnels. Light ground forces—elite infantry, SOF, and engineering forces—supported by heavy ground forces—tanks and artillery—worked together methodically at the tactical level to uncover and blow up these tunnels, assisted at the operational level by continuous IAF strikes against infrastructure targets related to the network. The diligent application of traditional CAW at the tactical and operational levels resulted in the destruction of most of the network of tunnels. With their mission complete, IDF ground forces withdrew from Gaza in early August, though cross-border fighting—that is, coalition rocket fire and IAF air strikes—would linger on intermittently until a firm cease-fire came into effect in late August.

Operation Protective Edge, like Operation Cast Lead, ended in a clear-cut Israeli victory. Diplomatically speaking, the Hamas-led coalition accepted a firm cease-fire, despite the fact that Israel had not made any political concessions whatsoever to the coalition. Furthermore, in the years since the operation, the Israel–Gaza border has been rather quiet in comparison to earlier years. Military speaking, thousands of insurgents/terrorists had been killed or wounded in the fighting, the coalition's military infrastructure throughout Gaza had been heavily damaged, its network of attack tunnels had been largely destroyed, and its rocket fire into Israel had been rendered impotent. Indeed, the IAF's antirocket interceptor system shot down 85–90 percent of the rockets it attempted to intercept, namely those rockets heading for inhabited areas of the country.[23] Most of the few Israeli civilian casualties in the war were inflicted by mortar fire into areas directly adjacent to the Israel–Gaza border, not by the thousands of rockets launched into the interior of the country.

Still, Operation Protective Edge was not quite the walkover of Operation Cast Lead. The Hamas-led coalition proved able to fire large numbers of rockets into Israel right up until the cease-fire took hold, its inability to inflict serious casualties or damage notwithstanding. Moreover, while IDF losses in men and machines amounted to only a small fraction of the losses inflicted upon the Hamas-led coalition during the war, they were by no means insignificant in themselves. In contrast to Operation Cast Lead, during which coalition forces often fell apart under even mild IDF pressure, they generally put up a stiff fight this time around.

Summary

The decline of IW as a surpassing threat to Israel over the past four decades, along with the concomitant rise of AW as such a threat over the same time interval, has exercised a profound influence on the IDF's approach to war. Offensive maneuver warfare, the dominant mode of warfare during the era of IW, has clearly taken a back seat to offensive attrition warfare, which has become the dominant mode of warfare in the era of AW. Rather than attempt to engage in swift and deep thrusts into its adversaries' rear areas to bring about their collapse, IDF ground maneuvers in all three of Israel's AWs have been shallow and slow-moving affairs intended to hammer the country's foes into submission. Firepower, especially from the air, has been favored over mobility.

Perhaps no development better exemplifies the switch from offensive maneuver warfare to offensive attrition warfare than the Israeli defense establishment's decision to build a comprehensive, multilayered integrated air defense system (IADS) to protect the country from rockets and missiles. Begun in the 1980s, as the Syrian and Iraqi ballistic missile programs gathered steam, and given further impetus in the 1990s as the Israeli home front came under ballistic missile attack during the 1991 Gulf War and as the Iranian ballistic missile program gathered steam, an antiballistic missile interceptor system was first deployed by the IDF in 2000. A short-range rocket interceptor system joined the Israeli IADS in 2011 in order to counter rocket attacks on the home front from Lebanon and Gaza. And a medium-range rocket/missile interceptor system designed to counter longer-range rockets and cruise missiles was deployed by the IDF in early 2017, completing the country's full-spectrum IADS. The reason that the Israeli defense establishment has countenanced the investment of so many resources in the creation of this state-of-the-art IADS lies in its belief that massive exchanges of standoff fire rather than massive ground maneuvers will continue to prevail in future rounds of war.

7

Between Wars

Combined Arms Warfare and Low-intensity Conflict

The State of Israel has historically divided its national security challenges into two broad categories: basic security challenges and current security challenges. The former category includes those challenges that have represented a threat to the fundamental integrity of the country: interstate war (IW) with Arab countries; asymmetrical war (AW) with Arab nonstate actors, namely Hizbullah and Hamas; and the potential use of weapons of mass destruction (WMD) against the Israeli home front by Arab states and Iran. The latter category consists of those challenges that have represented a threat not to the fundamental integrity of the country but rather to the welfare of its citizenry: cross-border infiltration and cross-border skirmishing; terrorism at home and abroad; and mass civil unrest and insurgency in Judea, Samaria, and Gaza. All of these current security challenges, as different from one another as they may be in their essence, fall under the general rubric of low-intensity conflict (LIC).

The Israel Defense Forces (IDF) has been kept quite busy waging LICs between wars. Indeed, it has been involved in one form or another of LIC across the entirety of Israel's history, with the exception of a few years from the late 1950s through the early 1960s, an unusually quiet period in the Arab–Israeli conflict. While the scope, intensity, and duration of these LICs has also varied throughout the years, the IDF has traditionally applied the principles of combined arms warfare (CAW) in its efforts to subdue its different adversaries. Moreover, the complexity of its approach to CAW in LIC has grown steadily over the decades.

Combined Arms Warfare and Low-intensity Conflict Up to the Six-Day War

The 1947–49 War of Independence concluded with signed armistice agreements between Israel and neighboring Arab states; however, these agreements did not lead to quiet along the country's frontiers. With the exception of the Israel–Lebanon border, which remained calm until the 1967 Six-Day War, all of Israel's frontiers began to burn shortly after the end of the War of Independence.[1] The Israel–Syria border witnessed frequent exchanges of fire, many of these episodes stemming directly from efforts by one or the other of the protagonists to alter the status quo on bits of territory claimed by both of them. The Israel–Jordan and Israel–Egypt frontiers, however, proved to be the most troublesome for Israel, as the vast majority of the civilian and military casualties incurred by the country between the War of Independence and the 1956 Sinai Campaign were suffered along these borders, where large-scale Arab infiltration triggered Israeli cross-border retaliation, which in turn led to repeated border skirmishes between Israel and its neighbors.

In the first years after the War of Independence, much of this infiltration was rather benign in character, as persons displaced by the war sought to reunite with family members and/or to retrieve abandoned property; however, some of this infiltration was clearly sinister in character, with insurgents/terrorists seeking to kill Israelis and destroy their property. Furthermore, the problem of armed infiltration became more acute with the passage of time. Indeed, hundreds of Israeli civilians and soldiers would die as a result of armed infiltration between the War of Independence and the Sinai Campaign.[2]

Initially, the Israeli defense establishment sought to handle infiltration through a combination of defensive measures, including mobile (foot, camel, and motorized) patrols and fortified border strongpoints/communities. It soon became apparent, however, that such measures alone would not suffice to stem the tide of armed infiltration. The country's long frontiers and lack of strategic depth, combined with the limited resources that it could devote to border security, meant that insurgents/terrorists could easily reach Israeli towns and villages to wreak havoc among their residents. The defense establishment resolved, therefore, that Israel must supplement its defensive posture with an offensive one in the form of retaliatory raids into Jordanian and Egyptian territory. The establishment reasoned that the raids would eventually coerce Jordan and Egypt into cracking down on cross-border infiltration once they themselves began to suffer costs.

During its first years, the policy of retaliatory raids turned out to be less than a grand success. The IDF itself admitted as much, acknowledging that considerably fewer than half of the raids achieved their objectives.[3] Generally executed by poorly trained and poorly commanded infantry forces unfamiliar with the principles of traditional CAW at the tactical level—in which small arms, machine guns, and explosives were often not employed in a mutually supportive manner—the early raids did not alleviate the problem of armed infiltration on Israel's central and southern borders.

From a narrow military point of view, the raids became much more effective after 1953, when they were carried out exclusively by well-trained and well-commanded parachute infantry and special operations forces (SOF) familiar with the basics of traditional CAW. These forces consistently displayed a high degree of skill in integrating small arms, machine guns, explosives, and, more and more, armored vehicles and heavier weapons in order to inflict serious losses on the Jordanian and Egyptian armies at much lower cost to themselves. Operation Black Arrow, an attack on a well-fortified Egyptian army camp in Gaza in early 1955, demonstrated the extent to which the IDF had mastered the intricacies of fire and movement.[4]

Still, from a broader national security point of view, the raids' impact on armed infiltration did not live up to the defense establishment's expectations. On the one hand, the losses suffered by the Jordanian army in these border skirmishes did push Jordan to adopt measures to clamp down on armed infiltration from its territory, spurring a decline in the scope and intensity of the problem on the Israel–Jordan border, at least for awhile. On the other hand, however, Egypt increased its support for armed infiltration into Israel as a consequence of raids against Egyptian army installations. By the mid-1950s, therefore, the Israeli defense establishment reckoned that the policy of retaliatory raids had run its course.[5] A war, according to the defense establishment, would be necessary to bring about a cessation of armed infiltration, not to mention the border skirmishes that stemmed from this phenomenon.

The Sinai Campaign did indeed bring about an almost complete cessation of armed infiltration for more than a decade. Not until after the Six-Day War would this phenomenon again become a real problem for Israel. The Sinai Campaign also brought about a notable lull in border skirmishing between Israel and its Arab neighbors until the mid-1960s. This relative calm was shattered by what would come to be known as the "Water War," a series of exchanges of fire across the Israel–Syria border in which the IDF specifically targeted a large-scale construction project intended to divert the Jordan River's waters from entering Israeli territory.[6] During these exchanges of fire, IDF tanks and artillery, as well as Israel Air Force (IAF) fighter-bombers, worked together not only to destroy Syrian engineering equipment, thereby putting an end to the water diversion project, but also to silence Syrian tanks and artillery firing into Israeli territory. The principles of traditional CAW at the tactical level, in other words, were in evidence during this round of hostilities. Border skirmishing between Israel and Syria would continue intermittently after this "war," eventually triggering a series of events that would culminate in the Six-Day War.

Combined Arms Warfare and Low-intensity Conflict Up to the Lebanon War

In one of the tragic ironies of Israeli history, the IDF's decisive victory in the Six-Day War actually sparked an intensification rather than diminution of hostilities in the Arab–Israeli conflict. Over the next few years, Israel would be drawn into two more IWs: the 1969–70 War of Attrition and the 1973 Yom Kippur War. Furthermore, a similar spike in LIC occurred during the years between the Six-Day War and the Yom Kippur War.

The conquest of Judea, Samaria, and Gaza brought a large and restless Palestinian population under Israeli administration. The Palestine Liberation Organization (PLO), which had been formed in 1964, took advantage of this situation to embark on an insurgency/terror campaign against Israel from both inside and outside of these territories.[7] The IDF again resorted to a combination of defensive and offensive measures to counter insurgent/terrorist forces. In Judea, Samaria, and Gaza, the defensive measures included electronic surveillance, mobile infantry patrols, observation posts, strongpoints, minefields, and fencing/barbed wire, all linked in a web based on the principles of traditional CAW. The offensive measures included search, seizure, and arrest raids, targeted attacks on insurgent/terrorist commanders, proactive ambushes based on intelligence derived from electronic and human sources, and intermittent raids against Arab states, mainly Jordan, until PLO forces were expelled from that country during its 1970 civil war, and, thereafter, Lebanon, where PLO forces subsequently took up residence.

Regardless of whether its actions were defensive or offensive in nature, the IDF functioned according to the principles of traditional CAW at the tactical level. Parachute infantry and SOF, supported by special border police and intelligence units, skillfully integrated machine guns, small arms, explosives, and, sometimes, heavier weapons in order to defeat PLO bands that formed inside Judea and Samaria or that attempted to slip into these territories from Jordan. Moreover, IAF helicopters were often used to transport parachute infantry and SOF in pursuit of these bands. By 1970, the IDF had defeated PLO forces in Judea and Samaria through the prudent application of traditional CAW, inflicting far more casualties on its adversary than it had suffered itself. And, by 1971, Gaza had been pacified as well, targeted attacks against insurgent/terrorist leaders playing a very prominent role in the IDF's successful counterinsurgency/counterterrorist campaign in this neck of the woods. Consequently, the PLO increasingly resorted to international terrorism, as Israeli/Jewish targets located abroad were far more vulnerable.

The IDF's proficient application of the principles of traditional CAW during this counterinsurgency/counterterrorist campaign is perhaps nowhere better illustrated than in its operations in Jordan and Lebanon. Two examples suffice to illustrate the point. In the spring of 1968, Operation Hell saw an all-arms IDF task force composed primarily of parachute infantry (some transported in armored vehicles and some transported in helicopters), tanks, and engineers, backed up by IAF fighter-bombers, cross the Jordan River to attack a heavily fortified PLO base at Karameh, drawing the Jordanian army into the fray. A fierce battle ensued that claimed the lives of more than 200 PLO insurgents and Jordanian soldiers, along with approximately 30 IDF troops.[8] Operation Remedy, a simultaneous diversionary attack on other PLO bases in Jordan, claimed the lives of another 75 or so PLO insurgents and Jordanian soldiers. Not a single Israeli parachute infantry or tank soldier perished in the latter operation. In the spring of 1973, Operation Spring of Youth saw a joint task force composed of parachute infantry and SOF, ferried to the Lebanese capital of Beirut aboard Israel Navy (IN) fast patrol boats (FPBs), attack a number of targets in and around the city with machine guns, small arms, and explosives.[9] Several dozen PLO insurgents, including three high-ranking commanders, were killed in these attacks at the cost of only two IDF troops. Exploiting their mastery of fire and movement, the parachute infantry and SOF swiftly destroyed their targets and withdrew from the area aboard IN FPBs and IAF helicopters before PLO and Lebanese army/police forces could organize a coherent defense.

Border skirmishing between Israel and its Arab neighbors between the Six-Day War and the Yom Kippur War was not limited to Jordan and Lebanon. The Israel–Syria and Israel–Egypt borders were also aflame during these years. Though Syria would not allow PLO forces to launch attacks from its own territory, it actively assisted the PLO's insurgency/terrorist campaign, both in the Middle East and in the international arena, which in turn sparked border skirmishes, as Israel sought to disrupt Syrian–PLO cooperation. A particularly notable instance of hostilities across the Israel–Syria frontier occurred in the early summer of 1970, when an IDF task force composed of tanks and engineers, supported by artillery and fighter-bombers, crossed the cease-fire line on the Golan in order to engage Syrian armor in a "day of battle," destroying 30–40 Syrian tanks in a textbook case of traditional CAW.[10] Along the Israel–Egypt frontier—the only Israeli frontier, incidentally, where the PLO's insurgency/terrorist campaign played no role in the border skirmishing—the fighting prior to the War of Attrition consisted primarily of exchanges of artillery fire across the Suez Canal; however, the IDF carried out a least one successful infantry raid into the Egyptian hinterland employing IAF helicopters to transport parachute infantry forces to their targets.

The Israel–Lebanon frontier remained Israel's only "active" border in respect of LIC after the Yom Kippur War, with clashes between IDF and PLO forces a common occurrence.[11] Similar to its practice in Judea, Samaria, and Gaza, the IDF employed both defensive and offensive measures in its effort to thwart PLO attacks on the Israeli home front. The defensive measures again included electronic surveillance, mobile patrols, observation posts, strongpoints, minefields, and fencing/barbed wire, all of which were woven together in a combined arms web. While generally effective at preventing infiltration, a number of PLO bands nevertheless managed to penetrate the country, perpetrating a series of grisly terror attacks throughout the 1970s.[12] The offensive measures consisted mainly of air strikes and artillery bombardments against PLO targets; however, the IDF also engaged in ground raids, often in tandem with air support, throughout the 1970s. By far the largest action of this kind, Operation Litani in 1978, saw an all-arms task force of tanks, mechanized and parachute infantry, artillery, and engineers, backed up by airpower, advance to the Litani River in southern Lebanon. Through the effective application of the principles of CAW, the IDF task force carved up some of the PLO forces in the area, while sending others fleeing northward toward Beirut. Still, despite the fact that the IDF inflicted heavy casualties on PLO forces in Lebanon during the 1970s, the threat to Israel posed by the PLO would not be eliminated until its expulsion from Lebanese territory as a result of the 1982 Lebanon War.

Combined Arms Warfare and Low-intensity Conflict Up to the New Millennium

From the Lebanon War to the turn of the twenty-first century, Israel confronted an old LIC challenge from a new adversary in Lebanon and a new LIC challenge from an old adversary in Judea, Samaria, and Gaza. No sooner had the PLO been expelled from Lebanon than a new Iranian-backed insurgent/terrorist group, which would eventually come to be known as Hizbullah, took its place. For the next 18 years, the IDF and Hizbullah would engage in hostilities in southern Lebanon.[13] To fight Hizbullah, the IDF relied on a mix of defensive and offensive measures. The former included electronic surveillance, mobile patrols, observation posts, strongpoints, fencing and barbed wire, and minefields, as well as cooperation with the South Lebanon Army (SLA), an Israeli-supported militia stationed in the area, all woven once again into a web based on the principles of traditional CAW. The latter included air strikes, artillery bombardments, raids by SOF, targeted attacks on Hizbullah commanders, as well as large-scale incursions by all-arms task forces on two occasions. During these large-scale incursions—Operations Accountability (1993) and Grapes of Wrath (1996)—IDF SOF worked in tandem with fighter-bombers and artillery in order to inflict casualties on Hizbullah personnel and to destroy the group's infrastructure in southern Lebanon in a display of traditional CAW.

Though it experienced some setbacks in southern Lebanon—such as the premature termination of Operation Grapes of Wrath when a cluster of errant artillery shells caused heavy civilian casualties, as well as a bungled raid by IN SOF that led to the loss of 13 soldiers—the IDF generally applied the principles of traditional CAW to good effect during the LIC in southern Lebanon.[14] Hizbullah suffered heavy losses among its personnel and heavy damage to its infrastructure over the course of the conflict. Nevertheless, the IDF was unable to defeat the group, whose capabilities grew steadily from the early 1980s through the late 1990s, partly as a result of ever-growing support in respect of foreign training and arms, mainly from Iran, and partly as a result of widespread ideological and logistical support from the Shi'ite populace of southern Lebanon. Indeed, Hizbullah proved able not only to fight IDF forces in Lebanon, but also eventually able to mount attacks against Israeli towns and villages in the north via rocket fire, as well as terror attacks against Israeli/Jewish targets abroad.[15] In the spring of 2000, therefore, Israel chose to withdraw unilaterally from southern Lebanon, no longer willing to suffer the costs of its presence there, primarily the steady trickle of IDF casualties produced by the hostilities. The Israeli defense establishment reasoned that a unilateral pull out would actually enhance the country's national security by restoring calm to the Israel–Lebanon frontier, a judgment that would later turn out to be tragically wrong.

While the conflict between Hizbullah and Israel raged in southern Lebanon, the IDF faced a second LIC challenge in Judea, Samaria, and Gaza from the late 1980s to the early 1990s in the form of the Intifada.[16] This LIC differed fundamentally from any other faced by Israel before or since in that the country's adversary was not a state or an insurgent/terrorist group. The adversary during the Intifada was "the street"—that is, the Palestinian populace of Judea, Samaria, and Gaza—whose restlessness under Israeli control exploded into a mass uprising in late 1987. This LIC also differed fundamentally from all other LICs faced by Israel in that it involved mainly rioting and vandalism rather than insurgency and terrorism. The country, to be sure, confronted "lone wolf" terror attacks that claimed Israeli lives, as well as organized Palestinian assassination squads, which mainly killed fellow Arabs accused (usually falsely) of collaborating with Israeli authorities, during the six years of the Intifada; however, IDF forces (together with border police forces) most commonly squared off against mobs hurling stones and firebombs, burning tires and other assorted debris, and/or destroying property.

To counter this LIC, the IDF resorted once more to a combination of defensive and offensive measures. The defensive measures included electronic surveillance, mobile patrols, observation posts, riot control operations, mass arrests/detentions (of rioters), checkpoints, and restrictions on movement (of the general population). The offensive measures included proactive intelligence-gathering operations to locate "wanted men" and the deployment of IDF undercover "Arab masquerader" units to arrest (in the vast majority of cases) or (if circumstances warranted) to kill (in rare cases) these men, most of whom were either uprising ringleaders or assassination squad members.[17]

Though the Intifada would not formally come to an end until the signing of the 1993 Oslo Accords between Israel and the PLO, the IDF had essentially got on top of the uprising by the beginning of the 1990s. Its assiduous (and sometimes quite harsh) application of defensive and offensive measures, coupled with the sheer exhaustion of the Arab street, resulted in a significant decline in Israeli–Palestinian violence during the last years of the Intifada. Intra-Palestinian violence, on the other hand, grew steadily during the same years, as the assassination squads picked up the pace of their campaign against fellow Arabs deemed politically or morally incorrect.

The practice of traditional CAW was not in evidence during the Intifada, as the IDF was basically employed in a law enforcement capacity. Still, intra-agency cooperation among IDF undercover forces, border police forces, and civilian security personnel (from the Shin Bet, the country's internal security service) to take down wanted men amounted to a form of CAW at the tactical level, albeit a non-traditional form. The use of different weapons—small arms (both lethal and nonlethal types), stun grenades, tear gas, odorous liquids, and gravel-throwing machines—during riot control operations can also be viewed as a form of CAW at the tactical level, again albeit an atypical form.

One particularly troublesome outcome of the Intifada for Israel was the emergence of the insurgent/terrorist group known as Hamas, which would quickly become a serious thorn in the country's side. As the Oslo peace process floundered during the mid and late 1990s, this group engaged in a sustained terror campaign of homicide bombings and kidnappings, taking a heavy toll on Israeli civilians and soldiers alike.[18] The destructiveness of this campaign notwithstanding, it was only a taste of what was in store for Israel in the early years of the twenty-first century.

Combined Arms Warfare and Low-intensity Conflict in the Twenty-first Century

In an effort to broker a final, comprehensive Israeli–Palestinian peace agreement, the United States invited both sides to a summit at Camp David in the summer of 2000. No agreement, however, emerged from the summit, as the Palestinian Authority (PA), the Palestinians' ruling body, rejected every attempt to reach a mutually acceptable compromise, preferring instead to fall back on violence in an effort to coerce Israel into accepting a "peace of the grave."[19] In the fall of 2000, therefore, PA-affiliated insurgent/terrorist groups such as Fatah, as well as Islamist insurgent/terrorist groups such as Hamas, began a massive preplanned insurgency/terror campaign against Israel, seizing on the visit of a prominent Israeli politician to the Temple Mount in Jerusalem as a pretext to initiate the violence.

The Second Intifada, also sometimes referred to as the al-Aqsa Intifada, which would last until 2005, at least insofar as concerned Judea and Samaria, bore very little resemblance to its predecessor.[20]

The Palestinian street played only a minor role in this round of LIC, and only during the first few months of hostilities. From the outset, insurgent/terrorist groups intent on bleeding Israel into submission dominated the Second Intifada. And not since the War of Independence did the Israeli home front suffer as much death and destruction. From 2000–5, more than 1,000 Israelis, most of them civilians, perished in a seemingly endless string of homicide bombings (often misleadingly referred to as "suicide bombings"), other bombings, mass shootings, stabbing sprees, and so forth. Thousands more were injured in this terror campaign.[21] Concomitantly, the insurgent/terrorist groups engaged in a full-blown insurgency against the IDF in Judea, Samaria, and Gaza.

Initially, the IDF (and the border police) relied primarily on defensive measures to counter this insurgency/terror campaign. These measures included electronic surveillance, mobile patrols, observation posts, checkpoints and roadblocks, as well as fencing and barbed wire, all stitched together in a web based on the principles of traditional CAW. Offensive measures, which included air strikes and ground incursions into PA-controlled territory, as well as targeted attacks on insurgent/terrorist commanders, were sporadic and limited in nature in an effort to keep the violence from escalating further. This mix of mostly defensive measures, however, did not produce the desired results, as Israeli casualties continued to mount during the first two years of the Second Intifada.

Consequently, in the spring of 2002, the IDF went on the offensive, carrying out a major sweep in Judea and Samaria in order to root out the insurgent/terrorist groups there. During Operation Defensive Shield, the IDF resorted to the principles of traditional CAW to smash these groups. Infantry-led task forces, backed up by armor and engineers, successfully dislodged these groups from their urban strongholds in Judea and Samaria, killing hundreds of insurgents/terrorists and capturing thousands more. Airpower and artillery were employed sparingly, if at all, in order to minimize collateral damage to the surrounding civilian population.

Operation Defensive Shield signaled the beginning of the end of the Second Intifada in Judea and Samaria. In its aftermath, the IDF (and border police) maintained steady offensive pressure on the remnants of these groups as part of a larger process known as "mowing the grass" (which, simply put, means cutting the adversary down to size as necessary).[22] This pressure sometimes took the form of targeted attacks against insurgent/terrorist commanders, but mostly involved search and seizure raids and arrest raids. Undercover IDF (and border police) units carried out many of the latter raids.[23] These raids incorporated the principles of traditional CAW, at least to the extent that they were based on the intimate cooperation between intelligence and field units. Simultaneously, Israel built a defensive security barrier along much of the length of its border with Judea and Samaria. This combination of offensive and defensive measures led to a steep decline in Israeli casualties over the following years, such that the Second Intifada had essentially petered out in Judea and Samaria by 2005. Since then, the area has generally been quiet, except for an extended wave of "lone wolf" stabbing, shooting, and (car) ramming attacks that began in the autumn of 2015, but had significantly abated by the spring of 2016.

Gaza has been a different story. In order to end friction with the Palestinian population there and to encourage the resumption of the Israeli–Palestinian peace process, the Israeli defense establishment decided to withdraw unilaterally from Gaza. The defense establishment hoped that a PA-controlled Gaza free of an Israeli presence would turn into a good neighbor; however, the defense establishment failed to foresee that Hamas would oust the PA from Gaza in a bloody coup in 2007. The Hamas takeover, in turn, led to a dramatic upsurge in border skirmishing between Israel and Gaza. Rocket fire into southern Israel, which had been a rather minor problem so long as the IDF remained in Gaza now became a major problem, thoroughly undermining the security of an ever-growing segment of Israel's population and eventually precipitating two AWs. Before and between these wars, the IDF attempted to disrupt the rocket fire through air strikes and/or artillery bombardments. These offensive measures were supplemented by a defensive measure in the form of the deployment of an antirocket interceptor system in 2011. One particularly effective application of CAW in a LIC situation occurred during Operation Pillar of Defense in 2012, when airpower and antirocket interceptor system teamed up to deliver a heavy blow to Hamas (and allied groups) at little cost to the IDF and Israel.[24]

Summary

The IDF has traditionally relied on a mix of defensive and offensive measures to wage LICs, frequently resorting to the practice of traditional CAW to counter the threat posed to Israel by insurgent/terrorist groups. In the country's early years, the burden of fighting LICs fell almost entirely on infantry forces, so the practice of traditional CAW in this kind of conflict was limited largely to those infantry forces that undertook retaliatory raids into Arab territory. In later years, the IDF integrated other arms—armor, artillery, engineers, and airpower—into its increasingly complex practice of traditional CAW in LICs, as demonstrated in Operations Hell, Litani, Accountability, and Grapes of Wrath. During the twenty-first century, the IDF's approach to traditional CAW in LICs, whether applied on the defensive or on the offensive, has added yet more layers of complexity. The cooperation among civilian and military intelligence units, SOF, and airpower during targeted attacks on insurgent/terrorist commanders, for example, has exhibited a seamlessness and effectiveness completely unobtainable in the past century.

8

"In Time of Peace, Conflict, or War"

Combined Arms Warfare and Special Operations

Special operations (SPEC OPS), according to a broadly echoed United States armed forces definition of the term, are:

> [O]perations conducted . . . to achieve military, political, economic, or psychological objectives by unconventional military means in hostile, denied, or politically sensitive areas. . . . Special operations differ from conventional operations in the degree of physical and political risk, operational techniques, modes of employment, independence from friendly support, and dependence on detailed operational intelligence and indigenous assets.[1]

And special operations forces (SOF), according to an equally broadly echoed Canadian armed forces definition of the term, are:

> [O]rganizations containing specially selected personnel that are organized, equipped[,] and trained to conduct high-risk, high[-]value . . . operations . . . by using special and unique operational methodologies . . . to achieve desired tactical, operational[,] and/or strategic effects in time of peace, conflict[,] or war.[2]

SPEC OPS, to translate military jargon into the vernacular, can take a wide variety of forms and, therefore, so can SOF. The latter can consist exclusively of ground, air, or naval personnel, or some combination thereof. Regardless of their form, however, SOF have typically relied upon combined arms warfare (CAW) to achieve their objectives, perhaps even more so than their conventional brethren. The reason is rather simple: CAW is a force multiplier, and SOF, by their very nature, are usually outnumbered and outgunned by their adversaries on their adversaries' home turf; therefore, SOF require every advantage that they can lay their hands on in order to ensure the success of their missions.[3]

The Origins of Israeli Special Operations

The history of Israeli SPEC OPS stretches back to the decade prior to the establishment of the State of Israel, to the days of the Special Night Squads (SNS), joint British–Jewish forces under the command of the charismatic British officer Orde Wingate, who would later become famous for his Second World War exploits in Ethiopia and Burma.[4] The administration of British Mandate Palestine tasked Wingate and his SNS with hunting down and eliminating marauding Arab gangs who were attacking British, Jewish, and Arab targets throughout the country during the 1936–39 Arab Rebellion. Wingate taught the SNS to engage in "fire and movement," integrating armored vehicles, machine guns, small arms, and improvised explosive devices (IEDs) into their repertoire. Relying upon up-to-date, actionable intelligence and employing CAW at the tactical level, the SNS carried out numerous effective night raids against Arab gangs, taking them by surprise and defeating them in their home villages, thereby playing an outsized role in countering the Arab rebellion.

Wingate and the SNS eventually became victims of their own success. With the end of the Arab Rebellion in 1939, the Palestine administration had Wingate, viewed with suspicion for his pro-Jewish sentiments, transferred out of the country and the SNS disbanded in order to curry favor with the Arab world as the Second World War loomed on the horizon. Once the war the got underway, however, Great Britain soon turned once more to the Haganah (Defense), the Palestine Jewish community's militia, for assistance in the Middle Eastern theater. Thus, with British assistance, the Palmach (an abbreviation for Strike Companies) arose as the Haganah's SOF.[5] Trained in CAW at the tactical level, especially for reconnaissance missions, Palmach forces gave a good account of themselves in Lebanon and Syria in various actions as part of British operations to defeat pro-German Vichy French forces in the region. By the time of the 1947–49 War of Independence, Palmach forces had essentially evolved into standard infantry brigades and were absorbed into the newly formed

Israel Defense Forces (IDF) as such.

Following independence, the IDF's first experience with SOF took the form of Unit 101, which was established in the early 1950s to counter Arab infiltration across Israel's borders and which was merged into the parachute infantry within a few months.[6] Integrating armored vehicles, machine guns, small arms, and IEDs into a fire and movement model similar to the one employed by the SNS, this force conducted numerous operations in Jordan, Egypt, and Syria from 1953–56. Its practice of CAW in nighttime raids paid handsome dividends at the tactical level. Operations such as Black Arrow and Elkayam (Egypt), Eli–Zait (Syria), and Shomron (Jordan) generally inflicted heavy casualties on the adversary at limited cost to the IDF.[7]

Nevertheless, the birth of modern Israeli SPEC OPS—in substance if not in spirit—springs most directly from the creation of Sayeret Matkal, the IDF General Staff's reconnaissance unit, which gradually developed a joint tactical SOF doctrine based on CAW in tandem with the first helicopter squadron of the Israel Air Force (IAF) beginning in the late 1950s.[8] This doctrine has been refined and expanded ever since those years, and it has been put into practice in literally hundreds—probably thousands—of SPEC OPS, large and small, carried out by the IDF over the succeeding decades, all of them employing CAW to one extent or another.

Combined Arms Warfare and Modern Israeli Special Operations

The nexus between modern Israeli SPEC OPS and CAW may be illustrated through an examination of four prominent operations: Spring of Youth, Jonathan, Show of Force, and Out of the Box. The first and third involved what nowadays is called "targeted killing," the second involved hostage rescue, and the fourth involved weapons of mass destruction (WMD). No claim is made here that these four operations constitute a representative sample of modern Israeli SPEC OPS. Rather, the claim made here is that, collectively speaking, they impart a strong sense of how the IDF has practiced CAW in the context of such operations.

Operation Spring of Youth

In response to the massacre of Israeli athletes at the 1972 Olympic Games, the Israeli government authorized a wide-ranging counterterrorism program to destroy Black September, a faction of the Palestine Liberation Organization (PLO), whose members had carried out the slaughter. Specifically, the government authorized the MOSSAD, Israel's foreign intelligence service, to eliminate Black September's leadership, most of which had found sanctuary in Europe. Over the course of months, the MOSSAD conducted a targeted-killing campaign throughout the continent, hunting down and "putting to sleep" much of Black September's leadership.

Part of that organization's leadership, however, had taken up residence in the Lebanese capital of Beirut. Rather than entrust their demise solely to the MOSSAD, the Israeli government approved an ambitious IDF SPEC OP in the spring of 1973 whose aim would be not only the deaths of the remaining Black September leaders, but also the destruction of various PLO targets throughout the city. Indeed, Operation Spring of Youth would be the largest and most complex SPEC OP to that point in Israeli history, involving IDF ground forces and the Israel Navy (IN) in central roles, as well as the MOSSAD and the IAF in supporting roles.

The operation had five separate components.[9] First, members of Sayeret Matkal would be responsible for killing three Black September leaders. Second, members of Sayeret Tzanchanim, the reconnaissance unit of the IDF's parachute infantry forces, would be responsible for destroying a command post belonging to the Popular Front for the Liberation of Palestine (PFLP), another faction of the PLO. Third, members of Shayetet 13, the IN's SOF, would be responsible for destroying a dual-purpose command post/arms factory belonging to Fatah, yet another faction of the PLO. Fourth, other members of Shayetet 13 would also be responsible for destroying a second ammunition dump belonging to Fatah. And, fifth, other members of Sayeret Tzanchanim would be responsible for destroying a third ammunition dump belonging to Fatah. All of the SOF teams would be transported to the vicinity of their respective targets in and around Beirut aboard IN fast patrol boats (FPBs), and would then be brought ashore aboard Zodiac boats crewed by members of Shayetet 13. Those SOF teams headed to targets in downtown Beirut would be driven to their destinations from the beach by MOSSAD operatives who had rented vehicles for this purpose. IAF helicopters would be on standby to remove any IDF dead and wounded from the field, while each of the five teams would withdraw in the same way that it arrived at its target.

In the event, all five of the components of Operation Spring of Youth went off like clockwork. The three Black September leaders, along with dozens of PLO gunmen, perished during the operation. Black September would never recover, largely ceasing to exist as an organization. The PFLP and Fatah command posts, the Fatah arms factory, and the Fatah ammunition dumps went up in flames as well. The IDF, in contrast, lost just two SOF troopers killed and three more wounded in the operation. The IAF evacuated the casualties by air, while the SOF teams were recovered and returned to Israel uneventfully by the IN.

Aside from the accurate and actionable intelligence provided by both A'MAN, the intelligence branch of the IDF, and the MOSSAD, the efficient practice of CAW at both the tactical and operational levels ensured that Operation Spring of Youth resulted in success. At the tactical level, each of the five SOF teams effectively integrated machine guns, small arms, rifle grenades, and explosives in the context of IDF fire and movement protocols in order to overcome their adversaries. And, at the operational level, cooperation among the SOF teams, the IN, the MOSSAD, and the IAF proved to be quite efficient, notwithstanding a hiccup here and there.

Operation Jonathan

In the summer of 1976, a mixed group of Arab and German terrorists, operating under the aegis of the PLO, hijacked an Air France airliner out of Athens filled with Israeli and Jewish passengers.[10] After a brief stopover in Libya, the airliner eventually flew to Entebbe Airport, Uganda. For its part, the Ugandan government, in solidarity with the PLO, had agreed prior to the hijacking to lend Uganda's support and protection to the hijackers, who demanded the release of terrorists in Israel and Europe as the price of freedom for the hostages.

Initially, the Israeli government sought to negotiate indirectly with the hijackers, as a hostage-rescue operation so far from Israeli territory—Entebbe was approximately 2,000 miles (3,200 km) from the nearest IAF base (in the southern Sinai)—appeared to be extremely impractical. Within a few days of the hijacking, however, IDF general headquarters (GHQ) had formulated a feasible hostage-rescue operation based on accurate and actionable intelligence collected by both A'MAN and the MOSSAD.

First known as Thunderbolt—and later renamed in honor of the fallen commander of Sayeret Matkal—the operation called for IAF transport aircraft to fly IDF SOF directly to Entebbe, where the latter would then assault the airport and free the hostages, who would in turn be flown home after a brief stopover in Kenya, which had secretly agreed to assist Israel in consequence of its poor relationship with Uganda. Specifically, Operation Jonathan called for four IAF transport aircraft to ferry elements of Sayeret Matkal, Sayaret Tzanchanim, and Sayaret Golani, the reconnaissance unit of the elite Golani Infantry Brigade, to Entebbe. Sayeret Matkal teams would be responsible for securing the airport's old terminal building, where the hostages were being held by a small group of terrorists. Sayaret Tzanchanim teams would be responsible for keeping the airport's runways open and for securing the airport's new terminal building, the latter in order to prevent the Ugandan army from intervening in the operation on behalf of the terrorists. And Sayeret Golani teams would be responsible for getting the liberated hostages to the IAF transport aircraft and for acting as a reserve ready to assist Sayeret Matkal and Sayeret Tzanchanim teams.

Operation Jonathan, like Operation Spring of Youth, went off like clockwork, notwithstanding a few hiccups here and there: no operation, after all, ever unfolds entirely according to plan. Flying at low altitude to avoid detection en route to Entebbe, the IAF transport aircraft delivered the SOF teams to the airport on time and in secret. Sayeret Matkal teams promptly stormed the old terminal building, killing all of the terrorists on scene and liberating the hostages. Sayeret Tzanchanim teams promptly took control of the airport's runways and seized the new terminal building, preventing a coordinated Ugandan army response to the Israeli assault. And Sayeret Golani teams quickly shepherded the liberated hostages to the IAF transport aircraft and provided fire support to the Sayeret Matkal and Sayeret Tzanchanim teams. In less than one hour, the approximately 100 hostages were in the air on their way to Israel via Kenya. In less than two hours, all IDF forces had withdrawn from Entebbe. Seven terrorists and several dozen Ugandan soldiers lay dead, scattered around the airport. At least eight Ugandan air force fighter-bombers had been destroyed on the tarmac (because they could have posed a threat to the outbound IAF transport aircraft), along with other assorted military equipment belonging to the Ugandan army. The IDF lost only one man killed in action, the aforementioned commander of Sayeret Matkal, and four hostages perished as well, three during the operation itself and one in hospital murdered later by Ugandan security personnel.

Operation Jonathan has gone down in history as one of the most audacious and successful SPEC OPS ever to be undertaken. The IDF's efficient practice of CAW at the tactical and operational levels constitutes one of the principal reasons for the triumph at Entebbe. At the tactical level, Israeli SOF teams effectively integrated armored vehicles, machine guns, small arms, and, where appropriate, rocket-propelled grenades (RPGs) and explosives to defeat their adversaries swiftly and decisively. Moreover, timely command and control (C^2) exercised on the ground (rather than from afar) allowed the different SOF teams to act in close coordination with each other. And, at the operational level, cooperation between the IAF and the SOF teams was as close to flawless as it is possible to achieve in such situations.

Operation Show of Force

Khalil al-Wizar, better known by his nom de guerre of Abu Jihad, was perhaps the PLO's leading terrorist mastermind, planning some of the worst atrocities carried out against Israel during the 1970s and 1980s. Not surprisingly, then, he had earned a well-deserved place near the top of the country's "hit list." His effort to escalate the level of violence during the first Intifada led the Israeli government to authorize the IDF to kill him at its earliest opportunity.[11]

Operation Show of Force, like Operation Jonathan, would take place far from Israel's shores, as Abu Jihad resided in Tunisia, some 1,500 miles (2,400 km) distant, where the PLO maintained its headquarters (HQ). And, like Operations Spring of Youth and Jonathan, this SPEC OP would also be quite complex, requiring not only SOF but also IN, IAF, and MOSSAD assets. Indeed, the IN would commit several FPBs, a submarine, and a disguised merchant ship to the operation, while the IAF would commit an airborne electronic warfare (AEW) aircraft, an airborne warning and control system (AWACS) aircraft, an airborne tanker, fighter-bombers, and a helicopter. For its part, the MOSSAD had a number of operatives on the ground in Tunisia to assist the SOF with the operation.

According to the plan of Operation Show of Force, teams from Sayeret Matkal and Shayetet 13 would be transported to the vicinity of Tunisia aboard IN FPBs, the former to execute the targeted killing of Abu Jihad and the latter to secure the beach where the SOF would come ashore. MOSSAD operatives would then guide the Sayeret Matkal team to the villa in the Tunisian capital of Tunis where al-Wazir was holed up. An IN submarine would first reconnoiter the beach to make sure that it was safe for the Sayeret Matkal and Shayetet 13 teams to proceed with the landing. A disguised merchant ship carrying an IAF helicopter would wait offshore with the FPBs, just in case the SOF teams on the ground needed to be reinforced with additional teams or to be extracted under fire. Overhead, meanwhile, IAF AEW and AWACS aircraft would monitor and, if necessary, disrupt Tunisian military communications, while fighter-bombers patrolled the area above the naval task force. C^2 of the entire operation would be exercised from a command post on one of the FPBs.

Like Operations Spring of Youth and Jonathan, Operation Show of Force unfolded essentially according to plan. In the spring of 1988, the designated Shayetet 13 team secured the "clean" beach, and the designated Sayeret Matkal team quickly followed ashore. With the assistance of the MOSSAD operatives waiting near the beach, the Sayeret Matkal team proceeded uneventfully to Abu Jihad's villa. The team then entered the home, eliminated Abu Jihad, and withdrew from the home, all in less than five minutes. The withdrawal of the SOF teams and MOSSAD operatives from Tunisia was smooth and unopposed, as was the withdrawal of IN and IAF assets from the Tunisian coast. The IDF suffered no losses in the operation.

In addition to the accurate and actionable intelligence furnished by A'MAN and the MOSSAD right up until the final moments before the commencement of the operation, Show of Force succeeded because of the efficient practice of CAW at the operational level. Cooperation among the SOF teams, the IN task force, the IAF assets, and the MOSSAD operatives during the operation was seamless.

Operation Out of the Box

In the spring of 2007, alarm bells went off inside the Israeli government when the MOSSAD discovered that Syria was well on its way toward completing a nuclear weapons production facility on its territory (at a place called al-Kibar in the northeast of the country) with North Korean assistance.[12] A subsequent deep-penetration reconnaissance SPEC OP carried out either by Sayeret Matkal or Shaldag, the elite reconnaissance unit of the IAF, or by a combination of the two, definitively confirmed the purpose of the installation, as soil samples collected at the site by the SOF and brought back to Israel revealed clear traces of radioactive particles associated with the fabrication of nuclear weapons.[13]

Once in possession of irrefutable evidence of Syria's intention to build a bomb, the Israeli government resolved to destroy the al-Kibar installation, as Israel could not live with the prospect of nuclear weapons in the hands of a mortal adversary. The Israeli government first approached the White House to determine whether the American administration would be willing to order the United States Air Force (USAF) to destroy the facility.[14] The White House, however, declined to take action, citing the country's unwillingness to risk the chance of a large-scale conflict with Syria while it was caught up in wars in Afghanistan and Iraq. The Israeli government, therefore, authorized the IAF to destroy the al-Kibar installation at the earliest opportunity.

Operation Out of the Box (which has also been referred to unofficially as Operation Orchard) would be a joint air–cyber affair. It called for IAF fighter-bombers to penetrate Syrian airspace and bomb the facility under the cover of an electronic screen, which would blind the Syrian integrated air defense system (IADS) to the approach of the Israeli aircraft. Specifically, cyber warriors from A'MAN's Unit 8200, the Israeli equivalent of the American National Security Agency (NSA), would take control of the IADS, ensuring that its operators did not detect any Israeli air activity in Syrian skies on the night of the attack.

In the summer of 2007, Operation Out of the Box was put into motion, and it unfolded precisely according to plan. How the software that gave Unit 8200 complete control of the Syrian IADS was introduced into the system—that is, whether remotely via a stealth drone that used the beam of a Syrian radar to download the software into the system or directly via SOF physically inserting the software into the system—remains a well-guarded secret. In either case, on the night of the attack, Unit 8200 had complete control of the IADS, and the Syrian operators saw exactly what Israel wanted them to see—nothing. Under the cover of this electronic screen, IAF fighter-bombers reached the vicinity of their target undetected, dropped a reported 17 tons of precision-guided munitions (PGMs) on to it, and made a clean getaway. The attack utterly destroyed the al-Kibar installation, leaving only the foundations of the facility intact.

Undoubtedly, the IAF could have reached and destroyed the target without the assistance of Unit 8200's cyber warriors. A fully alert Syrian IADS could not have blocked the attack. The effective cooperation between (de facto) cyber and air SOF, however, just as undoubtedly rendered it easier to execute successfully Operation Out of the Box than would have been the case had CAW at the tactical level not been practiced in this instance.

Summary

A widely endorsed view in international defense circles holds that:

> [The] increased presence and unprecedented expansion of SOF is largely a result of the "new" kinds of conflicts that have emerged in the twenty-first century, comprising non-state and state-based

> actors using terrorist and guerilla tactics. . . . Classic force-on-force, state-based military models of conflict have come to be seen as less useful when confronting highly mobile and ever-changing adversaries in asymmetric conflicts. . . . SOF are seen to be ideally suited to counter-insurgency and missions in asymmetric conflicts. . .[15]

The Israeli defense establishment shares this view. Consequently, the IDF has increased its reliance on SOF throughout the past two decades. Concomitantly, IDF SOF have also continuously refined the practice of CAW at the tactical and operational levels in the context of SPEC OPS, incorporating new elements, such as cyber warfare, into their repertoire. The further transformation of CAW at the tactical and operational levels in the context of SPEC OPS is a process that appears set to continue within the IDF SOF community in order to improve that community's capability to defend Israel's national security interests.

Conclusion
The Benefits of Combined Arms Warfare

The State of Israel emerged in the mid-twentieth century as a fragile agricultural polity. In the seven decades since its establishment, however, it has thoroughly transformed itself into a vigorous industrial power. During that stretch of time, Israel has become the strongest state in the Middle East, and a medium power by world standards. In certain areas—scientific advancement, technological prowess, and military capabilities, for example—it ranks among the leading global polities. Moreover, the country's citizenry is among the most long-lived and satisfied in the world.[1] These facts speak to the effectiveness of the Israel Defense Forces (IDF) in defending the state across its history, occasional military blunders and setbacks notwithstanding. Israel, of course, could not have reached these heights in the absence of a robust IDF able to defend the country against its adversaries.

Indeed, the IDF emerged victorious in five of the six interstate wars (IWs) in which it has been a participant to date: the 1947–49 War of Independence; the 1956 Sinai Campaign; the 1967 Six-Day War; the 1973 Yom Kippur War; and the 1982 Lebanon War. Only the 1969–70 War of Attrition ended less conclusively, in a stalemate along the Suez Canal. Similarly, the IDF emerged victorious in all three of the asymmetrical wars (AWs) in which it has been a participant to date: the 2006 Second Lebanon War, the 2008–9 Gaza War (Operation Cast Lead), and the 2014 Gaza War (Operation Protective Edge), even if the first of these wars did not result in as thorough an Israeli triumph as the IDF had intended prior to hostilities. Furthermore, the IDF has done a creditable job of protecting Israel during the country's seemingly endless low-intensity conflicts (LICs). Despite the fact that this type of conflict often does not lend itself to clear-cut victory, the IDF has registered a number of unambiguous triumphs in this arena, such as the defeat of Palestinian insurgents/terrorists after the Six-Day War and during the 2000–5 Second Intifada. In other cases, such as the border skirmishing that took place during the country's early years and during the first Intifada, the IDF contained its adversaries until related military and/or diplomatic developments brought these LICs to an end. And, in the arena of special operations (SPEC OPS), the IDF's overall performance has been quite efficient.

The explanation for the IDF's success in IWs, AWs, LICs, and SPEC OPS is surely complex, involving the interaction of many variables; however, the IDF's embrace of the principles of traditional combined arms warfare (CAW) at both the tactical and operational levels is surely a very important component of this explanation. Regardless of whether it has been involved in IWs, AWs, LICs or SPEC OPS, the IDF has generally performed at its best when committed to traditional CAW. Witness, for example, its effective performance during the Six-Day War on the Sinai, Judean–Samarian, and Golan fronts; the later stages of the Yom Kippur War on the Sinai and Golan fronts; Operation Cast Lead; and the Second Intifada in Judea and Samaria from Operation Defensive Shield onward. Moreover, when the IDF has abandoned traditional CAW, it has often gotten itself into trouble. Witness, for example, the IDF's far less effective performance during the War of Attrition; the opening stage of the Yom Kippur War on the Sinai front; and the Second Lebanon War.

While an IW may be a low-probability event for Israel in the foreseeable future as a result of its peace treaties with Egypt and Jordan, as well as the chaos that has overtaken Iraq and Syria, an AW or a LIC is a high-probability event, as insurgent/terrorist groups like Hizbullah and Hamas who thrive in the current Middle Eastern environment continue to adhere firmly to their objective of harming the country. Consequently, the IDF must be prepared to engage in traditional CAW to defend Israel against these threats. If the past is any guide to the future, the IDF will remain up to this task.

Glossary of Technical Terms

Airborne Electronic Warfare (AEW) Platform: an aircraft specially equipped to engage in electronic warfare (EW) tasks, such as jamming radar and/or communications systems, in order to impede an adversary's ability to conduct air, naval, and/or ground operations.

Airborne Warning and Control System (AWACS) Platform: an aircraft specially equipped to serve as a command and control (C^2) post in the sky, monitoring the location and disposition of an adversary's air, naval, and/or ground forces, as well as directing friendly air, naval, and/or ground forces to their targets.

Air Superiority: the attainment of control of the airspace above the battlefield.

Air Supremacy: a more definitive form of air superiority.

Antiaircraft Artillery (AAA): artillery that is optimized to shoot down aircraft. This artillery may be guided by radar or optical systems, and it may be deployed in single units or in batteries made up of multiple units.

Antiradiation Missile (ARM): a missile designed exclusively to destroy an adversary's radar units.

Arm: a specific combat branch of the armed forces, such as armor, artillery, infantry, engineers, intelligence, air force, navy, and so on. The term is sometimes also used to refer to noncombat branches, such as logistics and personnel.

Armored Personnel Carrier (APC): an armored vehicle designed for use by the infantry as a fighting platform on the battlefield; however, it may also be used simply to transport infantry to or around the battlefield rather than as a fighting platform.

Asymmetrical War (AW): a sustained high-intensity conflict between (or among) a state (or states), on the one hand, and an insurgent/terrorist group (or groups), on the other hand.

Attrition Warfare: a mode of warfare that favors firepower over mobility.

Close Air Support (CAS): air strikes mounted against adversarial ground forces in close proximity to friendly ground forces in order to support the latter.

Cluster Bomb Unit (CBU): a bomb that disperses a large number of bomblets over a wide area. Such bombs are often employed against personnel, but they may also be used against tanks, APCs, artillery, and the like.

Combined Arms Warfare (CAW): a mode of warfare that is based on the premise that better results on the battlefield will be achieved by relying on a mix of different arms rather than on a single arm. Both balanced (symmetrical) and unbalanced (asymmetrical) versions of CAW have been tested on the battlefield.

Command and Control (C^2): the principles and process behind the management of friendly ground, air, and/or naval forces in order to achieve a specified objective (or set of objectives) on the battlefield.

Command, Control, Communications, and Intelligence (C^3I): a more sophisticated version of C^2.

Defensive Warfare: responding to the initiative of an adversary on the battlefield.

Electronic Countermeasures (ECM): see electronic warfare.

Electronic Warfare (EW): the employment of electronic systems by friendly forces to disrupt the ability of adversarial forces to conduct operations on the battlefield.

Fast Patrol Boat (FPB): a type of small, fast, and maneuverable warship often fitted with antiship missiles in order to allow it to engage larger naval vessels.

Force Structure: the way in which armed forces are organized and equipped for battle.

Half-track: a type of APC that has both wheels (in the front) and tank tracks (in the back).

Improvised Explosive Device (IED): a homemade bomb cobbled together from various spare parts and explosive substances.

Integrated Air Defense System (IADS): an air defense system that combines SAM and AAA batteries, as well as their associated fire-control radars, into a seamless web in an effort to deny an opposing air force the ability to attack the area the IADS is intended to defend.

Interdiction: air strikes mounted against adversarial ground forces before they come into contact with friendly ground forces in order to support the latter.

Interstate War (IW): a sustained high-intensity conflict between (or among) states.

Israeli Defense Establishment: the IDF and the defense ministry in respect of issues related exclusively to the development of force structure and war-fighting doctrine; the civilian government, in cooperation with the IDF, in respect of decisions to undertake military action.

Low-intensity conflict (LIC): persistent low-level hostilities between (or among) a state (or states), on the one hand, and a state (or states), an insurgent/terrorist group (or groups), a hostile population (or populations), or some combination of these entities, on the other hand.

Maneuver Warfare: a mode of warfare that favors mobility over firepower.

Mechanized Infantry: infantry that is mounted on and fights from APCs. For the sake of terminological simplicity, this term also covers armored infantry, a special category of mechanized infantry that trains specifically to support an army's tank forces and is directly incorporated into those forces.

Nonmechanized Infantry: infantry that fights primarily on foot. Such infantry is sometimes also referred to as "motorized" in those instances when it is transported to and around the battlefield in armored vehicles and/or trucks. Nevertheless, because this type of infantry typically dismounts before going into action, nonmechanized infantry is a more suitable term for it.

Offensive Warfare: taking the initiative against an adversary on the battlefield.

Operational Level of Warfare: the principles according to which larger military formations (i.e., divisions and armies) are organized for and employed in battle.

Order of Battle (ORBAT): the entire roster of equipment and units possessed by an army, or a circumscribed roster of equipment and units deployed by it in a particular battle.

Precision-guided Munition (PGM): a bomb, missile, or shell that is guided to a target by one of several means, including radar, television camera, laser beam, and ground positioning system (GPS) coordinates.

Revolution in Military Affairs (RMA): a way of war that emphasizes the advantages to be derived from the application of advanced technologies on the battlefield.

Rocket-propelled Grenade (RPG): an explosive projectile for use against fixed and mobile targets fired from either a dedicated launcher or an assault rifle.

Runway Destruction Bomb (RDB): a bomb designed exclusively to destroy air base runways.

Special Operation (SPEC OP): a sensitive, high-risk operation, often carried out deep inside an adversary's territory, by specially trained and equipped forces.

Special Operations Force (SOF): a force trained and equipped to carry out sensitive, high-risk operations, often deep inside an adversary's territory.

Surface-to-air Missile (SAM): a missile that is optimized to shoot down an aircraft. It is typically deployed as part of a battery that consists of several missile launchers and a fire-control radar unit. A single soldier can operate small, shoulder-launched missiles, which are typically heat seeking.

Surface-to-surface Missile (SSM): a missile that is optimized to destroy ground targets. This category of weapon includes short-range missiles, medium-range missiles, and long-range (or ballistic) missiles.

Tactical Level of Warfare: the principles according to which smaller military formations (i.e., squads through brigades) are organized for and employed in battle.

Unmanned Aerial Vehicle (UAV): a drone that is flown remotely by a pilot from a ground station. Historically, UAVs have been deployed primarily in order to gather intelligence about an adversary; however, over the past couple of decades, they have been employed increasingly to launch attacks against an adversary.

War-fighting Doctrine: the principles that govern how armed forces are to be deployed and utilized on the battlefield.

Weapons of Mass Destruction (WMD): weapons that are capable of inflicting a massive number of casualties. Chemical, biological, and nuclear weapons fall into this category, while conventional weapons are traditionally exempted from it.

Notes

Introduction

1 IWs are defined as sustained high-intensity conflicts between (or among) states.
2 AWs are defined as sustained high-intensity conflicts between (or among) states, on the one hand, and insurgent/terrorist groups, on the other hand.
3 LICs are defined as persistent low-level hostilities between (or among), states, on the one hand, and states, insurgent/terrorist groups, hostile populations, or some combination of these entities, on the other hand.
4 SPEC OPS are defined as sensitive high-risk operations that are often carried out deep inside an adversary's territory.
5 Offensive warfare is defined as taking the initiative against an adversary on the battlefield, while defensive warfare is defined as responding to the initiative of an adversary. Maneuver warfare is defined as an approach to battle that favors mobility over firepower, while attrition warfare is defined as an approach that favors firepower over mobility.
6 Though the line separating the tactical from the operational level of warfare has always been rather fluid, the former generally refers to the organization and employment of smaller military formations (i.e., generally squads through brigades), while the latter generally refers to the organization and employment of larger military formations (i.e., generally divisions and armies). For more on the distinction between the two, see Edward N. Luttwak, *Strategy: The Logic of War and Peace* (Cambridge, MA: Harvard University Press, 1987), pp. 82–112.
7 United States Army. Department of Army Headquarters, *Army Doctrine Reference Publication No. 3–0: Unified Land Operations* (Washington, D.C.: Department of the Army Headquarters, 2012), p. 1–14. For a jargon-free review of the concept, see Jonathan M. House, *Combined Arms Warfare in the Twentieth Century* (Lawrence, KS: University Press of Kansas, 2001), pp. 1–11. CAW can also, of course, take place within a single arm—for example, when an infantry force integrates mortars, machine guns, small arms, and explosives.

1 The War of Independence and the Sinai Campaign

1 For an overview of the Jewish community's efforts at self-defense prior to the establishment of the State of Israel, see Ze'ev Schiff, *A History of the*

Israeli Army: 1874 to the Present, Rev. ed. (New York: Macmillan Publishing Company, 1985), pp. 1–20.

2 An authoritative account of the histories of the Haganah, the Irgun, and Lehi, with an emphasis on the activities of the dissident organizations, can be found in Bruce Hoffman, *Anonymous Soldiers: The Struggle for Israel, 1917–1947* (New York: Knopf, 2015).

3 The history of the Jewish Brigade is presented in Morris Beckman, *The Jewish Brigade: An Army with Two Masters, 1944–45* (New York: De Capo Press, 1998).

4 Authoritative accounts of the intercommunal phase of the War of Independence can be found in Chaim Herzog, *The Arab–Israeli Wars: War and Peace in the Middle East from the War of Independence Through Lebanon*, Rev. ed. (New York: Vintage Books, 1984), pp. 17–45, and Benny Morris, *1948: A History of the First Arab–Israeli War* (New Haven: Yale University Press, 2008), pp. 75–179. On the balance of forces, see Herzog, pp. 17–24.

5 On Operation Nachshon and Plan D, see Herzog, *The Arab–Israeli Wars*, pp. 29–38, and Morris, *1948*, pp. 113–179.

6 Authoritative accounts of the interstate phase of the War of Independence can be found in Herzog, *The Arab–Israeli Wars*, pp. 46–104; Morris, *1948*, pp. 180–374; and Kenneth M. Pollack, *Arabs at War: Military Effectiveness, 1948–2001* (Lincoln, NE: University of Nebraska Press, 2002), pp. 15–24, 149–154, 269–282, and 448–455.

7 For the balance of forces on the eve of the interstate phase of the War of Independence, see Herzog, *The Arab–Israeli Wars*, pp. 47–48, and Pollack, *Arabs at War*, pp. 15–16, 150, 270–271, and 448.

8 The disastrous engagement at Latrun is chronicled in Herzog, *The Arab–Israeli Wars*, pp. 63–68; Morris, *1948*, pp. 219–231; and Pollack, *Arabs at War*, pp. 276–282.

9 Operations Yoav and Horev are traced in Herzog, *The Arab–Israeli Wars*, pp. 91–104, and Morris, *1948*, pp. 320–372.

10 For estimates of Israeli and Arab losses during the War of Independence, see Morris, *1948*, pp. 406–407.

11 Airpower's contribution to the Israeli war effort in the War of Independence is traced in Eliezer Cohen, *Israel's Best Defense: The First Full Story of the Israeli Air Force* (New York: Orion Books, 1993), pp. 1–60, and Ehud Yonay, *No Margin for Error: The Making of the Israeli Air Force* (New York: Pantheon, 1993), pp. 3–72.

12 The ATF would be absorbed into the IAF after the War of Independence.

13 The naval dimension of the Israeli war effort in the War of Independence is traced in Klaus Mommsen, *60 Years Israel Navy* (Norderstedt: Books on Demand, 2011), pp. 19–42.

14 On the pre-Sinai Campaign debate within the Israeli defense establishment over the merits of traditional CAW, see F. H. Toase, "The Israeli Experience of Armoured Warfare," pp. 163–164, in J. P. Harris and

F. H. Toase (eds.), *Armoured Warfare* (New York: St. Martin's Press, 1990).

15 For the Israeli–Egyptian balance of forces on the eve of the war, see Yagil Henkin, *The 1956 Suez War and the New World Order in the Middle East: Exodus in Reverse* (Lanham, MD: Lexington Books, 2015), pp. 114–115, and Pollack, *Arabs at War*, p. 31.

16 Authoritative accounts of Operation Kadesh 2 can be found in Herzog, *The Arab–Israeli Wars*, pp. 114–138; Henkin, *The 1956 Suez War*, pp. 119–127, 135–156, 159–191, 197, and 254; and Pollack, *Arabs at War*, pp. 32–40.

17 The engagement at the Mitla Pass is chronicled in Henkin, *The 1956 Suez War*, pp. 119–127 and pp. 159–169; Herzog, *The Arab–Israeli Wars*, pp. 117–123; and Pollack, *Arabs at War*, p. 32.

18 The engagement at Abu Ageila/Um Katef is chronicled in Henkin, *The 1956 Suez War*, pp. 135–156; Herzog, *The Arab–Israeli Wars*, pp. 123–129; and Pollack, *Arabs at War*, pp. 32–35.

19 The engagement at Rafah (Gaza) is chronicled in Henkin, *The 1956 Suez War*, pp. 169–175; Herzog, *The Arab–Israeli Wars*, pp. 129–134; and Pollack, *Arabs at War*, pp. 36–38.

20 The engagement at Sharm el-Sheikh is chronicled in Henkin, *The 1956 Suez War*, pp. 176–191, and Herzog, *The Arab–Israeli Wars*, pp. 134–138.

21 For estimates of Israeli and Egyptian losses, see Henkin, *The 1956 Suez War*, p. 254; Pollack, *Arabs at War*, p. 39; and Toase, "The Israeli Experience," p. 166.

22 Most historians of the war believe the actual figure is much closer to 1,000 than 3,000.

23 Cohen, *Israel's Best Defense*, p. 131, and Pollack, *Arabs at War*, p. 40.

24 Cohen, *Israel's Best Defense*, pp. 119–120, 121–122, 125–142 and 147, and Herzog, *The Arab–Israeli Wars*, pp. 120 and 127.

25 Cohen, *Israel's Best Defense*, pp. 106–108, and Henkin, *The 1956 Suez War*, pp. 119–121.

26 Cohen, *Israel's Best Defense*, pp. 111–113; Henkin, *The 1956 Suez War*, pp. 121–122; and Herzog, *The Arab–Israeli Wars*, pp. 117–118.

27 Herzog, *The Arab–Israeli Wars*, p. 136.

28 Henkin, *The 1956 Suez War*, p. 197, and Herzog, *The Arab–Israeli Wars*, p. 138.

2 The Six-Day War

1 Quoted in Michael B. Oren, *Six Days of War: June 1967 and the Making of the Modern Middle East* (New York: Oxford University Press, 2002), p. 306.

2 Richard E. Simpkin, *Tank Warfare: An Analysis of Soviet and NATO Tank Philosophy* (London: Brassey's, 1979), p. 36.

3 Kenneth M. Pollack, *Arabs at War: Military Effectiveness, 1948–1991* (Lincoln, NE: University of Nebraska Press, 2002), p. 326.

4 Quoted in Steven Pressfield, *The Lion's Gate: On the Front Lines of the Six Day War* (New York: Penguin, 2014), p. 151.
5 Jonathan M. House, *Combined Arms Warfare in the Twentieth Century* (Lawrence, KS: University Press of Kansas, 2001), p. 230.
6 On the fighter-bomber buildup between 1956–67, see Eliezer Cohen, *Israel's Best Defense: The First Full Story of the Israel Air Force* (New York: Orion Books, 1993), pp. 105 and 194. On the RDB, see ibid., p. 195. On the tank buildup, see F. H. Toase, "The Israeli Experience of Armoured Warfare," pp. 163 and 167, in J. P. Harris and F. H. Toase (eds.), *Armoured Warfare* (New York: St. Martin's Press, 1990).
7 House, *Combined Arms Warfare*, p. 230.
8 Ibid.
9 For the IDF ORBAT, see Eric Hammel, *Six Days in June: How Israel Won the 1967 Arab–Israeli War* (New York: Scribner's, 1992), pp. 428–430.
10 On the Egyptian army's plan to strike first, see Oren, *Six Days of War*, pp. 92–97.
11 For the Egyptian army's ORBAT, see Hammel, *Six Days in June*, p. 432.
12 Informed descriptions of the fighting on the Sinai front can be found in Simon Dunstan, *The Six Day War 1967: Sinai*, Vol. 1 (Oxford: Osprey Publishing, 2009), pp 40–87; Hammel, *Six Days in June*, pp. 172–279; Oren, *Six Days of War*, pp. 178–182, 200–202, 211–216, 247–250, 256, and 258–259; and Pollack, *Arabs at War*, pp. 64–74.
13 For the numbers of troops and tanks deployed by both the IDF and the Egyptian army, see Dunstan, *The Six Day War*, Vol. 1, p. 26, and Pollack, *Arabs at War*, pp. 58–59.
14 The break-in battle in the northern sector is chronicled in Dunstan, *The Six Day War*, pp. 40–55; Hammel, *Six Days in June*, pp. 172–218; Oren, *Six Days of War*, pp. 178–181; and Pollack, *Arabs at War*, pp. 64–67.
15 The break-in battle in the southern sector is chronicled in Dunstan, *The Six Day War*, Vol. 1, pp. 61–69; Hammel, *Six Days in June*, pp. 228–242; Oren, *Six Days of War*, pp. 181–182 and 201; and Pollack, *Arabs at War*, pp. 67–70.
16 The break-in battle in the central sector is chronicled in Dunstan, *The Six Day War*, Vol. 1, pp. 55–57 and 60–61; Hammel, *Six Days in June*, pp. 219–227; Oren, *Six Day of War*, pp. 211–212; and Pollack, *Arabs at War*, p. 71.
17 Oren, *Six Days of War*, p. 214, and Pollack, *Arabs at War*, p. 71.
18 The pursuit phase of the IDF offensive in the Sinai is chronicled in Dunstan, *The Six Day War*, Vol. 1, pp. 71–87; Hammel, *Six Days in June*, pp. 243–279; Oren, *Six Days of War*, pp. 247–250, 256, and 258–259; and Pollack, *Arabs at War*, pp. 71–74.
19 For estimates of Egyptian and Israeli losses in the Six-Day War, see Dunstan, *The Six Day War*, Vol. 1, p. 88; Hammel, *Six Days in June*, p. 279; Oren, *Six Days of War*, pp. 305–306; and Pollack, *Arabs at War*, p. 74.

20 On the Israeli diplomatic overture, see Oren, *Six Days of War*, p. 184.

21 The IDF offensive in Judea and Samaria is chronicled in Simon Dunstan, *The Six Day War 1967: Jordan and Syria*, Vol. 2 (Oxford, Osprey Publishing, 2009), pp. 33–64; Hammel, *Six Days in June*, pp. 283–383; Oren, *Six Days of War*, pp. 182–196, 202–207, 218–228, and 240–247; and Pollack, *Arabs at War*, pp. 299–315.

22 For the numbers of troops and tanks deployed by both the IDF and the Jordanian army, see Dunstan, *The Six Day War*, Vol. 2, pp. 26–29, and Pollack, *Arabs at War*, pp. 293–294.

23 For estimates of Jordanian and Israeli losses in the Six-Day War, see Dunstan, *The Six Day War*, Vol. 2, p. 64; Hammel, *Six Days in June*, p. 383; Oren, *Six Days of War*, pp. 305–306; and Pollack, *Arabs at War*, p. 315.

24 Developments on the Golan front from June 5–8 are reviewed in Oren, *Six Days of War*, pp. 228–231 and 260–262, and Pollack, *Arabs at War*, pp. 462–463.

25 The IDF offensive on the Golan is chronicled in Dunstan, *The Six Day War*, Vol. 2, pp. 68–88; Hammel, *Six Days in June*, pp. 387–424; Oren, *Six Days of War*, pp. 280–284, 292–294, and 300–303; and Pollack, *Arabs at War*, pp. 463–468.

26 For the numbers of troops and tanks deployed by both the IDF and the Syrian army, see Dunstan, *The Six Day War*, Vol. 2, pp. 28–32, and Pollack, *Arabs at War*, pp. 461–462.

27 For estimates of Syrian and Israeli losses in the Six-Day War, see Hammel, *Six Days in June*, p. 424; Oren, *Six Days of War*, pp. 305–306; and Pollack, *Arabs at War*, p. 468.

28 A comprehensive and authoritative account of Operation Focus, based on official documents and interviews with participating IAF personnel, can be found in Shlomo Aloni, *The June 1967 Six-Day War: Operation Focus*, Vol. A (Bat Hefer: Isradecal, 2008).

29 Each IAF aircraft involved in an air base attack made a single bombing run in order to disable a runway, followed by three strafing runs to destroy aircraft and smash infrastructure. E-mail communication, August 10, 2008, with Lieutenant Colonel (Ret.) Amos Cohen, who flew two air base attack sorties on June 5.

30 For estimates of the damage suffered by the Egyptian air force during Operation Focus, see Aloni, *The June 1967 Six-Day War*, pp. 179–180, and Oren, *Six Days of War*, p. 176.

31 For estimates of the damage suffered by the Jordanian, Syrian, and Iraqi air forces, see Aloni, *The June 1967 Six-Day War*, pp. 179–180.

32 Ibid., pp. 54 and 179.

33 This figure, as well as all subsequent figures on CAS and interdiction sorties, comes from the IAF Historical Branch. See the table in Kenneth M. Pollack, "Air Power in the Six-Day War," *The Journal of Strategic Studies*, Vol. 28, No. 3 (June 2005), pp. 471–503. The table is on p. 478.

34 On the IAF's limited impact with respect to the destruction of tanks, see the report by the Joint Technical Coordinating Group for Munitions Effectiveness, *Special Report: Survey of Combat Damage to Tanks*, Vol. I–III (Washington, D.C.: Defense Intelligence Agency, 1970). Also see Pollack, "Air Power," pp. 483–485.

35 For the IAF's logistical contributions to the Israeli offensives in the Sinai and on the Golan, see Martin van Creveld, "Israel: Maneuver Warfare, Air Power, and Logistics," pp. 166 and 168, in Martin van Creveld, Steven L. Canby, and Kenneth S. Brower, *Air Power and Maneuver Warfare* (Maxwell Air Force Base, AL: Air University Press, 1994).

36 The most authoritative account of this tragic and contentious incident can be found in A. Jay Cristol, *The Liberty Incident Revealed: The Definitive Account of the Israeli Attack on the U. S. Navy Spy Ship* (Annapolis, MD: Naval Institute Press, 2013).

37 On the capture of Sharm el-Sheikh, see Hammel, *Six Days in June*, pp. 251–253, and Oren, *Six Days of War*, pp. 247–248.

38 House, *Combined Arms Warfare*, p. 230.

3 The War of Attrition

1 Yaacov Bar-Siman-Tov, *The Israeli–Egyptian War of Attrition, 1969–1970: A Case Study of Limited Local War* (New York: Columbia University Press, 1980), p. 45, and Jonathan Shimshoni, *Israel and Conventional Deterrence: Border Warfare from 1953 to 1970* (Ithaca, NY: Cornell University Press, 1988), pp. 138–141.

2 Amos Amir, *Fire in the Sky: Flying in Defence of Israel* (South Yorkshire: Pen and Sword Books, 2005), pp. 46–56.

3 For a description of the Bar Lev Line, see Simon Dunstan, *Israeli Fortifications of the October War 1973* (Oxford: Osprey Publishing, 2008), pp. 16–19.

4 The events of the War of Attrition are chronicled in Bar-Siman-Tov, *The Israeli–Egyptian War of Attrition*, pp. 43–185; Chaim Herzog, *The Arab–Israeli Wars: War and Peace in the Middle East from the War of Independence through Lebanon*, Rev. ed. (New York: Vintage Books, 1984), pp. 195–202 and 207–219; and Shimshoni, *Israel and Conventional Deterrence*, pp. 123–211.

5 Bar-Siman-Tov, *The Israeli–Egyptian War of Attrition*, pp. 69–70.

6 Ibid., p. 87.

7 Shlomo Aloni, *Israeli Mirage and Nesher Aces* (Oxford: Osprey Publishing, 2004), pp. 82–83.

8 Bar-Siman-Tov, *The Israeli–Egyptian War of Attrition*, pp. 90–102.

9 Shlomo Aloni, *Israeli F-4 Phantom II Aces* (Oxford: Osprey Publishing, 2004), p. 86, and Aloni, *Israeli Mirage and Nesher Aces*, p. 83.

10 In-depth treatments of this bombing campaign may be found in Bar-Siman-Tov, *The Israeli–Egyptian War of Attrition*, pp. 117–173; Yaacov Bar-Siman-Tov, "The Myth of Strategic Bombing: Israeli Deep-

Penetration Air Raids in the War of Attrition, 1969–1970," *Journal of Contemporary History*, Vol. 19, No. 3 (July 1984), pp. 549–570; and Avi Shlaim and Raymond Tanter, "Decision, Process, Choice, and Consequences: Israel's Deep-Penetration Bombing in Egypt, 1970," *World Politics*, Vol. 30, No. 4 (July 1978), pp. 483–516.

11 For the number of deep-penetration air raids, see Bar-Siman-Tov, *The Israeli–Egyptian War of Attrition*, p. 132. For the number of sorties, see the description of the air force's part in the War of Attrition on the official IAF web site, <www.iaf.org.il>.

12 Bar-Simon-Tov, *The Israeli–Egyptian War of Attrition*, pp. 134–135.

13 Jon D. Glassman, *Arms for the Arabs: The Soviet Union and War in the Middle East* (Baltimore: Johns Hopkins University Press, 1975), p. 75, and Efraim Karsh, *The Cautious Bear: Soviet Military Engagement in Middle Eastern Wars in the Post-1967 Era* (Boulder, CO: Westview Press, 1987), pp. 16, 64–65, and 77. The most thorough treatment of Soviet involvement in the War of Attrition to date is Isabella Ginor and Gideon Remez, *The Soviet–Israeli War 1967–1973: The USSR's Military Intervention in the Egyptian–Israeli Conflict* (New York: Oxford University Press, 2017).

14 Shlomo Aloni, *Israeli F-4 Phantom II Aces*, p. 86, and Aloni, *Israeli Mirage and Nesher Aces*, pp. 83–84.

15 For IAF losses, see Eliot A. Cohen and John Gooch, *Military Misfortunes: The Anatomy of Failure in War* (New York: The Free Press, 1990), p. 99.

16 For estimates of Israeli and Egyptian manpower losses in the War of Attrition, see Dunstan, *Israeli Fortifications*, pp. 35–36.

17 W. Andrew Terrill, "The Nature and Value of Commando Operations During the Egyptian–Israeli War of Attrition," *Small Wars & Insurgencies*, Vol. 8, No. 3 (Autumn 1997), pp. 16–34, contains an overview of these IDF infantry raids. For a list of the raids during the second stage of the War of Attrition, see the chart in Bar-Siman-Tov, *The Israeli–Egyptian War of Attrition*, pp. 100–101.

18 An exhaustive account of the Green Island raid is found in Ze'ev Almog, *Flotilla 13: Israeli Naval Commandos in the Red Sea, 1967–1973* (Annapolis: Naval Institute Press, 2010), pp. 31–99.

19 On the sinking of the torpedo boats and the joint mechanized infantry–tank raid, see Bar-Siman-Tov, *The Israeli–Egyptian War of Attrition*, pp. 99–101; Herzog, *The Arab–Israeli Wars*, pp. 211–212; and Terrill, "The Nature and Value of Commando Operations," pp. 26–27.

20 The raid on the radar installation is addressed in Bar-Siman-Tov, *The Israeli–Egyptian War of Attrition*, pp. 100–101, and Terrill, "The Nature and Value of Commando Operations," pp. 27–28.

21 The raid on Shadwan Island is summarized in Herzog, *The Arab–Israeli Wars*, p. 214, and Terrill, "The Nature and Value of Commando Operations," p. 29.

22 Quoted in Bar-Siman-Tov, *The Israeli–Egyptian War of Attrition*, pp. 140–141.

23 On this strike, see Ehud Yonay, *No Margin for Error: The Making of the Israeli Air Force* (New York: Pantheon Books, 1993), pp. 305–306.

4 The Yom Kippur War

1 Anthony H. Cordesman and Abraham R. Wagner, *The Lessons of Modern War: The Arab–Israeli Conflicts, 1973–1989*, Vol. 1 (Boulder, CO: Westview Press, 1990), p. 54.
2 Robert S. Bolia, "Overreliance on Technology in Warfare: The Yom Kippur War as a Case Study," *Parameters*, Vol. 34, No. 2 (Summer 2004), pp. 51–52.
3 Emanuel Sakal, *Soldier in the Sinai: A General's Account of the Yom Kippur War* (Lexington, KY: University Press of Kentucky, 2014), p. 305.
4 For a distinction between the operational and tactical levels of warfare, see Edward N. Luttwak, *Strategy: The Logic of War and Peace* (Cambridge, MA: Harvard University Press, 1987), pp. 82–112.
5 On the tank buildup, see Sakal, *Soldier in the Sinai*, p. 98. On the aircraft buildup, see Central Intelligence Agency. Directorate of Intelligence, *The 1973 Arab–Israeli War: Overview and Analysis of the Conflict* (Langley, VA: Central Intelligence Agency, 1975), p. 28, and John F. Kreis, *Air Warfare and Air Base Air Defense* (Washington, D.C.: United States Government Printing Office, 1988), pp. 308 and 322. The CIA report was declassified in mid-2012. On the growth of the IAF's striking power, see Martin van Creveld, "Israel: Maneuver Warfare, Air Power, and Logistics," pp. 153–192, in Martin Van Creveld, Steven L. Canby, and Kenneth Brower, *Air Power and Maneuver Warfare* (Maxwell Air Force Base, AL: Air University Press, 1994).
6 Sakal, *Soldier in the Sinai*, pp. 326–327.
7 Cordesman and Wagner, *The Lessons of Modern War*, p. 55, and Sakal, *Soldier in the Sinai*, pp. 100–102.
8 Sakal, *Soldier in the Sinai*, p. 98.
9 Cordesman and Wagner, *The Lessons of Modern War*, pp. 103–104, and Sakal, *Soldier in the Sinai*, p. 102.
10 For extensive treatments of the assumptions underlying "the concept," see Dani Asher, *The Egyptian Strategy for the Yom Kippur War: An Analysis* (Jefferson, NC: McFarland, 2009); Uri Bar-Joseph, *The Watchman Fell Asleep: The Surprise of Yom Kippur and Its Sources* (Albany: SUNY Press, 2005); and Aryeh Shalev, *Israel's Intelligence Assessment Before the Yom Kippur War: Disentangling Deception and Distraction* (Brighton: Sussex Academic Press, 2010).
11 The Yom Kippur War has not lacked for chroniclers over the years. Among the many general histories of the war, two of the best are: Chaim Herzog, *The War of Atonement: October, 1973* (Boston: Little, Brown and Co., 1975), and Abraham Rabinovich, *The Yom Kippur War: The Epic Encounter that Transformed the Middle East* (New York: Schocken, 2004).
12 For the force ratios on the Golan, see Dani Asher (ed.), *Inside Israel's*

Northern Command: The Yom Kippur War on the Syrian Border (Lexington, KY: University Press of Kentucky, 2016), pp. 39–53 and 77–109; Directorate of Intelligence, *The 1973 Arab–Israeli War*, p. 53; Herzog, *The War of Atonement*, pp. 60–63; and Rabinovich, *The Yom Kippur War*, p. 142.

13 For the force ratios in the Sinai, see Directorate of Intelligence, *The 1973 Arab–Israeli War*, pp. 74–78; Herzog, *The War of Atonement*, pp. 150–151; Rabinovich, *The Yom Kippur War*, p. 55; and Sakal, *Soldier in the Sinai*, p. 77.

14 Asher, *Inside Israel's Northern Command*, pp. 39–53 and 77–109; Directorate of Intelligence, *The 1973 Arab–Israeli War*, pp. 52–53; Herzog, *The War of Atonement*, pp. 60–63; and Rabinovich, *The Yom Kippur War*, p. 142.

15 The IDF's defensive stand against the Syrian onslaught is traced in Asher, *Inside Israel's Northern Command*, pp. 127–323; Directorate of Intelligence, *The 1973 Arab–Israeli War*, pp. 54–64; Herzog, *The War of Atonement*, pp. 78–127; and Rabinovich, *The Yom Kippur War*, pp. 142–175 and 187–216.

16 On the role and conduct of the IDF's Golan strongpoints, see Asher, *Inside Israel's Northern Command*, pp. 127–323.

17 Ori Orr, *These Are My Brothers: A Dramatic Story of Heroism During the Yom Kippur War* (Jerusalem: Contento de Semrik, 2013), furnishes a first-hand account of the rapid mobilization and deployment of IDF reserve forces on the Golan front.

18 Asher, *Inside Israel's Northern Command*, pp. 537–573, describes the contributions of artillery and engineering forces, respectively, on the Golan front.

19 Ibid., pp. 575–599 describe the IAF's contribution on the Golan front.

20 For the IDF's offensive into Syria and the fighting until the war's end, see Asher, *Inside Israel's Northern Command*, pp. 341–478; Directorate of Intelligence, *The 1973 Arab–Israeli War*, pp. 64–68; Herzog, *The War of Atonement*, pp. 128–145; and Rabinovich, *The Yom Kippur War*, pp. 284–318.

21 Directorate of Intelligence, *The 1973 Arab–Israeli War*, p. 65.

22 On the recapture of the Mount Hermon strongpoint, see Asher, *Inside Israel's Northern Command*, pp. 427–478; Herzog, *The War of Atonement*, pp. 143–144; and Rabinovich, *The Yom Kippur War*, pp. 446–457.

23 Herzog, *The War of Atonement*, p. 151; Rabinovich, *The Yom Kippur War*, p. 55; and Sakal, *Soldier in the Sinai*, pp. 467–470.

24 On tank losses on the southern front during the first day of war, see Rabinovich, *The Yom Kippur War*, p. 233, and Sakal, *Soldier in the Sinai*, p. 174.

25 For the October 8 counterattack, see John J. McGrath, "Sinai 1973: Israeli Maneuver Organization and the Battle of the Chinese Farm," pp. 67–68, in John J. McGrath (ed.), *An Army at War: Change in the Midst of Conflict*

(Fort Leavenworth, KS: U.S. Army Command and General Staff College, 2005) and, especially, Sakal, *Soldier in the Sinai*, pp. 208–284.

26 McGrath, "Sinai 1973," p. 65, and Sakal, *Soldier in the Sinai*, p. 246.

27 Sakal, *Soldier in the Sinai*, pp. 208–284.

28 McGrath, "Sinai 1973," p. 69, and Sakal, *Soldier in the Sinai*, p. 300.

29 Directorate of Intelligence, *The 1973 Arab–Israeli War*, p. 89.

30 On the battle and the losses, see Directorate of Intelligence, *The 1973 Arab–Israeli War*, pp. 89–90; Herzog, *The War of Atonement*, pp. 205–207; Rabinovich, *The Yom Kippur War*, pp. 353–355; and Sakal, *Soldier in the Sinai*, pp. 302–304.

31 Treatments of the crossing operation can be found in Directorate of Intelligence, *The 1973 Arab–Israeli War*, pp. 91–97; Herzog, *The War of Atonement*, pp. 208–230; McGrath, "Sinai 1973," pp. 63–109; and Rabinovich, *The Yom Kippur War*, pp. 358–399.

32 For the IDF advance into Egypt, see Directorate of Intelligence, *The 1973 Arab–Israeli War*, pp. 97–99; Herzog, *The War of Atonement*, pp. 231–250; and Rabinovich, *The Yom Kippur War*, pp. 416–477.

33 For the IAF's prewar thinking, see Rabinovich, *The Yom Kippur War*, pp. 32–34, and Sakal, *Soldier in the Sinai*, pp. 328–333.

34 On American pressure to refrain from a preemptive strike, see the October 6, 1973 telegram from Secretary of State Henry Kissinger to President Richard Nixon in Middle East War Memos and Miscellaneous (October 1–October 17), National Security Council (NSC) File, Nixon Presidential Materials, Box 664, National Archives, and the October 7, 1973 memorandum of conversation between Kissinger and Israeli Ambassador to the United States Simcha Dinitz in Records of Henry Kissinger, 1973–1977, Record Group 59, State Department Records, Box 25 (Category C 1973 Arab–Israeli War), National Archives. Also see Michael Brecher, *Decisions in Crisis: Israel, 1967 and 1973* (Berkeley: University of California Press, 1980), pp. 177–178, 187–188, and 197–201.

35 Steven J. Rosen and Martin Indyk, "The Temptation to Pre-empt in a Fifth Arab–Israeli War," *Orbis*, Vol. 20, No. 3 (Summer 1976), pp. 271–272.

36 John R. Carter, *Airpower and the Cult of the Offensive* (Maxwell Air Force Base, AL: Air University Press, 1998), pp. 52–65.

37 In a conversation with the author on June 25, 2007 at the Israel Air Force Center (IAFC), Herzliya, Israel, Brigadier General (Ret.) Itzhak Amitay, who flew many combat missions on the southern front during the Yom Kippur War, confirmed that this operation had made progress before the IDF GHQ terminated it. For summaries of this operation, see Rabinovich, *The Yom Kippur War*, pp. 175–180, and Sakal, *Soldier in the Sinai*, pp. 338–342.

38 Rabinovich, *The Yom Kippur War*, pp. 178–179, and Sakal, *Soldier in the Sinai*, pp. 338-342.

39 For a summary and assessment of IAF activity in the Yom Kippur War,

see David Rodman, *Sword and Shield of Zion: The Israel Air Force in the Arab–Israeli Conflict, 1948–2012* (Brighton: Sussex Academic Press, 2013), pp. 31–38.

40 Sakal, *Soldier in the Sinai*, pp. 384–387.

41 Cordesman and Wagner, *The Lessons of Modern War*, pp. 94–95, and Sakal, *Soldier in the Sinai*, pp. 396–398.

42 For the IN war effort in the Mediterranean Sea arena, see Abraham Rabinovich, *The Boats of Cherbourg: The Navy that Stole Its Own Boats and Revolutionized Naval Warfare* (New York: Henry Holt, 1988).

43 For the IN war effort in the Red Sea arena, see Ze'ev Almog, *Flotilla 13: Israeli Naval Commandos in the Red Sea, 1967–1973* (Annapolis, MD: Naval Institute Press, 2010).

44 For estimates of aircraft losses during the Yom Kippur War, see Cordesman and Wagner, *The Lessons of Modern War*, p. 89. For estimates of tank and artillery losses, see the chart presented by the first commander of the United States Army's Training and Doctrine (TRADOC) Command in "Implications of the Middle East War on U.S. Army Tactics, Doctrine, and Systems," p. 79, in Richard M. Swain, Donald L. Gilmore, and Carolyn D. Conway, *Selected Papers of General William E. DePuy: First Commander, U.S. Army Training and Doctrine Command, 1 July 1973* (Fort Leavenworth, KS: U.S. Army Command and General Staff College, 1994). And, for estimates of casualties, see Ze'ev Schiff, *A History of the Israeli Army: 1874 to the Present*, Rev. ed. (New York: Macmillan, 1985), p. 228.

45 William E. DePuy, "Letter to Senator John C. Culver from General DePuy, 12 May 1975," p. 166, in Swain, Gilmore, and Conway, *Selected Papers of General William E. DePuy*.

5 The Lebanon War

1 A synopsis of the Agranat Commission's purpose and findings can be found at <www.knesset.gov.il>.

2 Anthony H. Cordesman and Abraham R. Wagner, *The Lessons of Modern War: The Arab–Israeli Conflicts, 1973–1989*, Vol. 1 (Boulder, CO: Westview Press, 1990), p. 110.

3 Ibid.

4 Cordesman and Wagner, *The Lessons of Modern War*, p. 111, and Martin van Creveld, *The Sword and the Olive: A Critical History of the Israeli Defense Force* (New York: PublicAffairs, 1998), p. 256.

5 van Creveld, *The Sword and the Olive*, p. 256.

6 David Elazar, "The Yom Kippur War: Military Lessons," p. 249, in Louis Williams (ed.), *Military Aspects of the Arab–Israeli Conflict* (Tel Aviv: University Publishing Projects, 1975).

7 Ibid., pp. 248–249.

8 Ibid., p. 249.

9 Cordesman and Wagner, *The Lessons of Modern War*, pp. 111–112.

10 Elazar, "The Yom Kippur War," p. 259.
11 For the balance of forces in the Lebanon War, see Cordesman and Wagner, *The Lessons of Modern War*, pp. 136–137; Trevor N. Dupuy and Paul Martell, *Flawed Victory: The Arab–Israeli Conflict and the 1982 War in Lebanon* (Fairfax, VA: HERO Books, 1986), pp. 86–94; and Kenneth M. Pollack, *Arabs at War: Military Effectiveness, 1948–1991* (Lincoln, NE: University of Nebraska Press, 2002), p. 525.
12 Hostilities between IDF and PLO forces in southern Lebanon are chronicled in Cordesman and Wagner, *The Lessons of Modern War*, pp. 136–144, and Dupuy and Martell, *Flawed Victory*, pp. 98–113 and 130–133.
13 Hostilities between IDF and Syrian army forces in southern Lebanon are chronicled in Cordesman and Wagner, *The Lessons of Modern War*, pp. 137–144; Dupuy and Martell, *Flawed Victory*, pp. 114–128; and Pollack, *Arabs at War*, pp. 526–539.
14 Events surrounding the siege of western Beirut are chronicled in Cordesman and Wagner, *The Lessons of Modern War*, pp. 144–147; Dupuy and Martell, *Flawed Victory*, pp. 134–140 and 148–163; and Pollack, *Arabs at War*, pp. 539–540.
15 For estimates of Israeli, PLO, and Syrian losses in men and machines, see Cordesman and Wagner, *The Lessons of Modern War*, p. 153; Dupuy and Martell, *Flawed Victory*, pp. 141–142; and Pollack, *Arabs at War*, p. 540.
16 On Operation Mole Cricket 19, see Cordesman and Wagner, *The Lessons of Modern War*, pp. 142 and 186–193; Dupuy and Martell, *Flawed Victory*, pp. 119–121; and Pollack, *Arabs at War*, pp. 532–534.
17 Cordesman and Wagner, *The Lessons of Modern War*, p. 153, and Pollack, *Arabs at War*, p. 540.
18 Cordesman and Wagner, *The Lessons of Modern War*, p. 202, and Dupuy and Martell, *Flawed Victory*, p. 120.
19 Cordesman and Wagner, *The Lessons of Modern War*, p. 202, and Pollack, *Arabs at War*, p. 540.
20 Cordesman and Wagner, *The Lessons of Modern War*, p. 153, and Dupuy and Martell, *Flawed Victory*, p. 145.
21 On the IAF's interdiction campaign, see Cordesman and Wagner, *The Lessons of Modern War*, pp. 205–206.
22 Cordesman and Wagner, *The Lessons of Modern War*, p. 143; Dupuy and Martell, *Flawed Victory*, p. 127; and Pollack, *Arabs at War*, p. 539.
23 On the IAF's CAS campaign, see Cordesman and Wagner, *The Lessons of Modern War*, pp. 203–205 and 210–213, and Dupuy and Martell, *Flawed Victory*, p. 146.
24 Cordesman and Wagner, *The Lessons of Modern War*, p. 212.
25 For the IN's participation in the Lebanon War, see Cordesman and Wagner, *The Lessons of Modern War*, pp. 137–138 and 216, and Dupuy and Martell, *Flawed Victory*, pp. 101, 106–108, and 146–147.

6 The Second Lebanon and Gaza Wars

1 That the IDF has embraced offensive attrition warfare does not imply that it has divested itself entirely of the capability to wage offensive maneuver warfare. The IDF still retains considerable tank and mechanized infantry forces that could carry a fight on the ground deep into Arab territory if such an undertaking were to be deemed necessary.

2 Eado Hecht and Eitan Shamir, *Medium-intensity Threats: The Case for Beefing-up IDF Ground Forces* (Ramat Gan: The Begin–Sadat Center for Strategic Studies, 2016), pp. 41–42, n. 11. Indeed, the IDF's arsenal today contains fewer tanks than at any time since the years just after the 1967 Six-Day War.

3 Ibid., p. 13.

4 Hecht and Shamir, *Medium-intensity Threats*, pp. 11 and 13, and Itamar Rabinovich and Itai Brun, *Israel Facing a New Middle East: In Search of a National Security Strategy* (Stanford: Hoover Institution Press, 2017), ch. 2.

5 Ibid.

6 The Second Lebanon War is chronicled in Amos Harel and Avi Isaacharoff, *34 Days: Israel, Hezbollah, and the War in Lebanon* (New York: Palgrave Macmillan, 2008).

7 The Israeli air campaign is traced in Itai Brun, "The Second Lebanon War, 2006," pp. 297–324, in John Andreas Olsen (ed.), *A History of Air Warfare* (Washington, D.C.: Potomac Books, 2010), and Benjamin S. Lambeth, *Air Operations in Israel's War Against Hezbollah: Learning from Lebanon and Getting It Right in Gaza* (Santa Monica: Rand Corporation, 2011).

8 Brun, "The Second Lebanon War," p. 317, and Lambeth, *Air Operations*, pp. 92–110.

9 Ibid.

10 The Israeli ground campaign is traced in Scott C. Farquhar (ed.), *Back to Basics: A Study of the Second Lebanon War and Operation Cast Lead* (Fort Leavenworth, KS: U.S. Army Combined Arms Center Combat Studies Institute Press, 2009); David Makovsky and Jeffrey White, *Lessons and Implications of the Israel–Hizballah War: A Preliminary Assessment* (Washington, D.C.: The Washington Institute for Near East Policy, 2006); and Matt M. Matthews, *We Were Caught Unprepared: The 2006 Hezbollah–Israel War* (Fort Leavenworth, KS: U.S. Army Combined Arms Center Combat Studies Institute Press, 2008).

11 Matthews, *We Were Caught Unprepared*, pp. 33–60. Also see Amir Rapaport, *The IDF and the Lessons of the Second Lebanon War* (Tel Aviv: The Begin–Sadat Center for Strategic Studies, 2010).

12 Matthews, *We Were Caught Unprepared*, pp. 54–55, and Rapaport, *The IDF*, p. 13.

13 On the IDF's post-Second Lebanon War reforms, see Matt M. Matthews, "Hard Lessons Learned," pp. 21–24, in Farquhar, *Back to Basics*; Abe F.

Merraro, "The Tactics of Operation Cast Lead," pp. 83–89, in Farquhar, *Back to Basics*; and Rapaport, *The IDF*, pp. 13–35.

14 The air and ground campaigns during Operation Cast Lead are chronicled in Amir Kulick, "'Lebanon Lite': Lessons from the Operation in Gaza and the Next Round Against Hizbollah," *Military and Strategic Affairs*, Vol. 1. No. 1 (April 2009), pp. 51–66; Lambeth, *Air Operations*, pp. 234–276; Matthews, "Hard Lessons Learned," pp. 24–34; and Merraro, "The Tactics of Operation Cast Lead," pp. 89–97.

15 An in-depth description and analysis of the initial Israeli air campaign is furnished in Lambeth, *Air Operations*, pp. 234–240.

16 Descriptions and analyses of the air–ground campaign can be found in Lambeth, *Air Operations*, pp. 240–259; Matthews, "Hard Lessons Learned," pp. 28–34; and Merraro, "The Tactics of Operation Cast Lead," pp. 89–97.

17 Matthews, "Hard Lessons Learned," p. 30; Merraro, "The Tactics of Operation Cast Lead," pp. 90 and 98; and Rapaport, *The IDF*, p. 23.

18 Quoted in Matthews, "Hard Lessons Learned," p. 30.

19 Kulick, "'Lebanon Lite,'" p. 55.

20 Gabriel Siboni, "Operations Cast Lead, Pillar of Defense, and Protective Edge: A Comparative Review," p. 32, in Anat Kurz and Shlomo Brom (eds.), *The Lessons of Operation Protective Edge* (Tel Aviv: Institute for National Security Studies, 2014).

21 The Database Desk, Incidents and Activists Database, *Operation 'Protective Edge': A Detailed Summary of Events* (Herzliya: International Institute for Counter-terrorism, 2014) is an in-depth review of the operation.

22 Ibid., pp. 2–4.

23 Uzi Rubin, *Israel's Air and Missile Defense During the 2014 Gaza War* (Ramat Gan: The Begin–Sadat Center for Strategic Studies, 2015), pp. 22–25.

7 Between Wars

1 For accounts of the hostilities along Israel's borders in the wake of the War of Independence, see Daniel Byman, *A High Price: The Triumphs and Failures of Israeli Counterterrorism* (New York: Oxford University Press, 2011), pp. 18–26; Gunther E. Rothenberg, "Israeli Defense Forces and Low-intensity Operations," pp. 55–59, in David Charters and Maurice Tugwell (eds.), *Armies in Low-intensity Conflict: A Comparative Analysis* (London: Brassey's Defense Publishers, 1989); and Jonathan Shimshoni, *Israel and Conventional Deterrence: Border Warfare from 1953 to 1970* (Ithaca, NY: Cornell University Press, 1988), pp. 34–122.

2 A (partial) accounting of Israeli casualties as a result of armed infiltration during these years is provided in Rothenberg, "Israeli Defense Forces," pp. 56 and 58. Ernest Stock, *Israel on the Road to Sinai, 1949–1956*

(Ithaca, NY: Cornell University Press, 1967), p. 71, offers additional data on casualties.

3 Rothenberg, "Israeli Defense Forces," p. 56.

4 On Operation Black Arrow, see Shimshoni, *Israel and Conventional Deterrence*, p. 79.

5 Moshe Dayan, *Diary of the Sinai Campaign* (New York: Schocken Books, 1967), p. 57.

6 On this round of border skirmishing, see Ami Gluska, "'The War over the Water' During the 1960s," pp. 109–131, in Mordechai Bar-On (ed.) *A Never-ending Conflict: A Guide to Israeli Military History* (Westport, CT: Praeger, 2004), and Avner Yaniv, *Deterrence Without the Bomb: The Politics of Israeli Strategy* (Lexington, MA: Lexington Books, 1987), pp. 106–107.

7 On the post-1967 Palestinian insurgency/terror campaign, see Byman, *A High Price*, pp. 34–54, and Rothenberg, "Israeli Defense Forces," pp. 61–67.

8 An in-depth description of Operation Hell can be found in Samuel M. Katz, *Follow Me!: A History of Israel's Military Elite* (London: Arms & Armour Press, 1989), pp. 93–95.

9 Ibid., pp. 112–116.

10 Alaric Searle, *Armoured Warfare: A Military, Political and Global History* (London: Bloomsbury, 2017), p. 161.

11 On the LIC across the Israel–Lebanon border from the Yom Kippur War to the Lebanon War, see Byman, *A High Price*, pp. 58–65, and Rothenberg, "Israeli Defense Forces," pp. 69–70.

12 For these attacks, see Byman, *A High Price*, p. 60, and Rothenberg, "Israeli Defense Forces," p. 69.

13 On the LIC between the IDF and Hizbullah, see Byman, *A High Price*, pp. 209–239, and Sergio Catignani, "Israeli Counter-insurgency Strategy and the Quest for Security in the Israeli–Lebanese Arena," pp. 67–89, in Clive Jones and Sergio Catignani (eds.), *Israel and Hizbollah: An Asymmetric Conflict in Historical and Comparative Perspective* (London: Routledge, 2009).

14 For the premature termination of Operation Grapes of Wrath, see Avi Kober, *Israel's Wars of Attrition: Attrition Challenges to Democratic States* (London: Routledge, 2009), p. 125. On the bungled IN raid, see Alex Fishman, "What Really Went Wrong in Botched 1997 Shayetet 13 Operation," *Yediot Ahronot*, June 21, 2017 (Internet edition). The latter article can be found on the newspaper's web site, <www.ynetnews.com>.

15 Hizbullah's two most notorious international terror attacks, both carried out with Iranian assistance, occurred in Argentina in the early 1990s, the targets being the Israeli embassy and a Jewish communal center.

16 On the Intifada, see Sergio Catignani, *Israeli Counter-insurgency and the* Intifadas*: Dilemmas of a Conventional Army* (London: Routledge, 2008), pp. 72–101, and Aryeh Shalev, *The Intifada: Causes and Effects* (Boulder, CO: Westview Press, 1991).

17 For a review of the masquerader units, see Stuart A. Cohen, *'Masqueraders' in the IDF, 1991–92: The Military Unit and the Public Debate* (Ramat Gan: Begin–Sadat Center for Strategic Studies, 1994).

18 The terror campaign waged by Hamas during these years is reviewed in Byman, *A High Price*, pp. 93–112.

19 On the agenda and modus operandi of the PA, see Efraim Karsh, *Arafat's War: The Man and His Battle for Israeli Conquest* (New York: Grove Press, 2003), pp. 156–199.

20 The Second Intifada is chronicled in Byman, A *High Price*, pp. 113–174, and Sergio Catignani, *Israeli Counter-insurgency*, pp. 102–141.

21 Israeli casualty figures during the Second Intifada are found in Catignani, *Israeli Counter-insurgency*, p. 105.

22 An in-depth description and analysis of this phenomenon can be found in Efraim Inbar and Eitan Shamir, "'Mowing the Grass': Israel's Strategy for Protracted Intractable Conflict," *Journal of Strategic Studies*, Vol. 37, No. 1 (February 2014), pp. 65–90.

23 Samuel M. Katz, *The Ghost Warriors: Inside Israel's Undercover War Against Suicide Terrorism* (New York: Berkley Caliber, 2016), examines how the border police's undercover unit has gone about conducting arrest raids.

24 On Operation Pillar of Defense, see Glen Segell, *Israel, Hamas and Gaza, 14–21 November 2012: Operation Pillar of Defence; Operation Stones of Shale; Operation Blue Sky* (London: The Institute of Security Policy, 2012).

8 "In Time of Peace, Conflict, or War"

1 Joint Chiefs of Staff, *Joint Publication 3–05 (Test): Doctrine for Joint Special Operations* (Washington, D.C.: Office of the Joint Chiefs of Staff, 1990), [n.p.]. Quoted in William H. McRaven, *Spec Ops: Case Studies in Special Operations Warfare, Theory and Practice* (Novato, CA: Presidio, 1995), p. 2, n. 1.

2 Government of Canada, *Canadian Special Operation Forces Command* (Ottawa: Department of National Defense, 2008), p. 7. Quoted in Bernd Horn, "The Evolution of SOF and the Rise of SOF Power," p. 25, n. 2, in Jessica Glicken Turnley, Kobi Michael, and Eyal Ben-Ari (eds.), *Special Operations Forces in the 21st Century: Perspectives from the Social Sciences* (London: Routledge, 2018).

3 For other variables identified as crucial to the success of SPEC OPS, see McRaven, *Spec Ops*, pp. 8–23. While the author does not explicitly identify CAW as one of the "principles" of success, he implicitly endorses its utilization in his model of SPEC OPS.

4 On Wingate and the SNS, see Simon Anglin, *Orde Wingate and the British Army, 1922–1944* (London: Pickering & Chatto, 2010), pp. 57–100, and John Bierman and Colin Smith, *Fire in the Night: Wingate of Burma, Ethiopia, and Zion* (New York: Random House, 1999), pp. 57–143.

5 On the Palmach and its operations, see Samuel M. Katz, *Follow Me!: A History of Israel's Military Elite* (London: Arms & Armour Press, 1989), pp. 16–21, and Mordechai Naor (ed.), *Ha'Haganah* (Tel Aviv: Ministry of Defense Publishing House, 1985), pp. 136–154.

6 Unit 101's and the parachute infantry's exploits are chronicled in Katz, *Follow Me!*, pp. 34–45.

7 The particulars of these operations are described in ibid., pp. 42–45.

8 Lior Brichta and Eyal Ben-Ari, "Organizational Entrepreneurship and Special Forces: the Israeli Helicopter Squadron and the General Staff Reconnaissance Unit (Sayaret Matkal)," pp. 212–225, in Turnley, Michael, and Ben-Ari, *Special Operations Forces in the 21st Century.*

9 The particulars of Operation Spring of Youth are described in Ronen Bergman, *Rise and Kill First: The Secret History of Israel's Targeted Assassinations* (New York: Random House, 2018), pp. 166–173, and Katz, *Follow Me!*, pp. 112–116.

10 The particulars of Operation Jonathan are described in Simon Dunstan, *Israel's Lightning Strike: The Raid on Entebbe* (Oxford: Osprey Publishing, 2009) and McRaven, *Spec Ops*, pp. 333–380.

11 The particulars of Operation Show of Force are described in Bergman, *Rise and Kill First*, pp. 315–323, and Michael Bar-Zohar and Nissim Mishal, *No Mission Is Impossible: The Death-defying Missions of the Israeli Special Forces* (New York: Ecco, 2015), pp. 257–265.

12 The particulars of Operation Out of the Box are described in Richard A. Clarke and Robert Knake, *Cyberwar: The Next Threat to National Security and What To Do About It* (New York: Ecco, 2010), pp. 1–8, and David Rodman, *Enforcing the Begin Doctrine: How Israel Stopped Iraq and Syria from Getting the Bomb* (Columbia, SC: CreateSpace, 2017), pp. 83–112.

13 Many of the specific details of the operation—such as the precise composition of the SOF that reconnoitered the Syrian nuclear installation—remain conjectural for the moment. The broader narrative of Operation Out of the Box, however, is not in doubt.

14 On Israel's overture to the United States in regard to the al-Kibar installation, see Warren Bass, *A Surprise Out of Zion?* (Santa Monica, CA: RAND Corporation, 2015), pp. 45–59.

15 Jessica Glicken Turnley, Eyal Ben-Ari, and Kobi Michael, "Special Operations Forces (SOF) and Social Sciences," p. 2, in Turnley, Michael, and Ben-Ari, *Special Operations Forces in the 21st Century.*

Conclusion

1 See Swiss–Israeli Association, "Life Expectancy in Israel on the Rise," *Israel Between the Lines*, June 1, 2017, and Zach Pontz, "And the Happiest Countries in the World Are . . .," *From the Grapevine*, March 16, 2016. These articles can be found at <www.israelbetweenthelines.com> and <www.fromthegrapevine.com>, respectively.

Bibliography

Articles

Bar-Siman-Tov, Yaacov, "The Myth of Strategic Bombing: Israeli Deep-Penetration Air Raids in the War of Attrition, 1969–1970," *Journal of Contemporary History*, Vol. 19, No. 3 (July 1984).

Bolia, Robert S., "Overreliance on Technology in Warfare: The Yom Kippur War as a Case Study," *Parameters*, Vol. 34, No. 2 (Summer 2004).

Fishman, Alex, "What Really Went Wrong in Botched 1997 Shayetet 13 Operation," *Yediot Ahronot*, June 21, 2017.

Inbar, Efraim and Eitan Shamir, "'Mowing the Grass': Israel's Strategy for Protracted Intractable Conflict," *Journal of Strategic Studies*, Vol. 37, No. 1 (February 2014).

Kulick, Amir, "'Lebanon Lite': Lessons from the Operation in Gaza and the Next Round Against Hizbollah," *Military and Strategic Affairs*, Vol. 1, No. 1 (April 2009).

Pollack, Kenneth M., "Air Power in the Six-Day War," *The Journal of Strategic Studies*, Vol. 28, No. 3 (June 2005).

Rosen, Steven J. and Martin Indyk, "The Temptation to Pre-empt in a Fifth Arab–Israeli War," *Orbis*, Vol. 20, No. 3 (Summer 1976).

Shlaim. Avi and Raymond Tanter, "Decision, Process, Choice, and Consequences: Israel's Deep-Penetration Bombing in Egypt, 1970," *World Politics*, Vol. 30, No. 4 (July 1978).

Terrill, W. Andrew, "The Nature and Value of Commando Operations During the Egyptian–Israeli War of Attrition," *Small Wars & Insurgencies*, Vol. 8, No. 3 (Autumn 1997).

Books and Studies

Almog, Ze'ev, *Flotilla 13: Israeli Naval Commandos in the Red Sea, 1967–1973* (Annapolis: Naval Institute Press, 2010).

Aloni, Shlomo, *Israeli F-4 Phantom II Aces* (Oxford: Osprey Publishing, 2004).

Aloni, Shlomo, *Israeli Mirage and Nesher Aces* (Oxford: Osprey Publishing, 2004).

Aloni, Shlomo, *The June 1967 Six-Day War: Operation Focus*, Vol. A (Bat Hefer: Isradecal, 2008).

Amir, Amos, *Fire in the Sky: Flying in Defence of Israel* (South Yorkshire: Pen and Sword Books, 2005).

Anglin, Simon, *Orde Wingate and the British Army, 1922–1944* (London: Pickering & Chatto, 2010).

Asher, Dani, *The Egyptian Strategy for the Yom Kippur War: An Analysis* (Jefferson, NC: McFarland, 2009).

Asher, Dani (ed.), *Inside Israel's Northern Command: The Yom Kippur War on the Syrian Border* (Lexington, KY: University Press of Kentucky, 2016).

Bar-Joseph, Uri, *The Watchman Fell Asleep: The Surprise of Yom Kippur and Its Sources* (Albany: SUNY Press, 2005).

Bar-On, Mordechai (ed.) *A Never-ending Conflict: A Guide to Israeli Military History* (Westport, CT: Praeger, 2004).

Bar-Siman-Tov, Yaacov, *The Israeli–Egyptian War of Attrition, 1969–1970: A Case Study of Limited Local War* (New York: Columbia University Press, 1980).

Bar-Zohar, Michael and Nissim Mishal, *No Mission Is Impossible: The Death-defying Missions of the Israeli Special Forces* (New York: Ecco, 2015).

Bass, Warren, *A Surprise Out of Zion?* (Santa Monica, CA: RAND Corporation, 2015).

Beckman, Morris, *The Jewish Brigade: An Army with Two Masters, 1944–45* (New York: De Capo Press, 1998).

Bergman, Ronen, *Rise and Kill First: The Secret History of Israel's Targeted Assassinations* (New York: Random House, 2018).

Bierman, John and Colin Smith, *Fire in the Night: Wingate of Burma, Ethiopia, and Zion* (New York: Random House, 1999).

Brecher, Michael, *Decisions in Crisis: Israel, 1967 and 1973* (Berkeley: University of California Press, 1980).

Byman, Daniel, *A High Price: The Triumphs and Failures of Israeli Counterterrorism* (New York: Oxford University Press, 2011).

Carter, John R., *Airpower and the Cult of the Offensive* (Maxwell Air Force Base, AL: Air University Press, 1998).

Catignani, Sergio, *Israeli Counter-insurgency and the* Intifadas*: Dilemmas of a Conventional Army* (London: Routledge, 2008).

Charters, David and Maurice Tugwell (eds.), *Armies in Low-intensity Conflict: A Comparative Analysis* (London: Brassey's Defense Publishers, 1989).

Clarke, Richard A. and Robert Knake, *Cyberwar: The Next Threat to National Security and What To Do About It* (New York: Ecco, 2010).

Cohen, Eliezer, *Israel's Best Defense: The First Full Story of the Israeli Air Force* (New York: Orion Books, 1993).

Cohen, Eliot A. and John Gooch, *Military Misfortunes: The Anatomy of Failure in War* (New York: The Free Press, 1990).

Cohen, Stuart A., *'Masqueraders' in the IDF, 1991–92: The Military Unit and the Public Debate* (Ramat Gan: Begin–Sadat Center for Strategic Studies, 1994).

Cordesman, Anthony H. and Abraham R. Wagner, *The Lessons of Modern*

War: The Arab–Israeli Conflicts, 1973–1989, Vol. 1 (Boulder, CO: Westview Press, 1990).

Cristol, A. Jay, *The Liberty Incident Revealed: The Definitive Account of the Israeli Attack on the U. S. Navy Spy Ship* (Annapolis, MD: Naval Institute Press, 2013).

Dunstan, Simon, *Israeli Fortifications of the October War 1973* (Oxford: Osprey Publishing, 2008).

Dunstan, Simon, *Israel's Lightning Strike: The Raid on Entebbe* (Oxford: Osprey Publishing, 2009).

Dunstan, Simon, *The Six Day War 1967: Sinai*, Vol. 1 (Oxford: Osprey Publishing, 2009).

Dunstan, Simon, *The Six Day War 1967: Jordan and Syria*, Vol. 2 (Oxford, Osprey Publishing, 2009).

Dupuy, Trevor N. and Paul Martell, *Flawed Victory: The Arab–Israeli Conflict and the 1982 War in Lebanon* (Fairfax, VA: HERO Books, 1986).

Farquhar, Scott C. (ed.), *Back to Basics: A Study of the Second Lebanon War and Operation Cast Lead* (Fort Leavenworth, KS: U.S. Army Combined Arms Center Combat Studies Institute Press, 2009).

Ginor, Isabella and Gideon Remez, *The Soviet–Israeli War 1967–1973: The USSR's Military Intervention in the Egyptian–Israeli Conflict* (New York: Oxford University Press, 2017).

Glassman, Jon D., *Arms for the Arabs: The Soviet Union and War in the Middle East* (Baltimore: Johns Hopkins University Press, 1975).

Hammel, Eric, *Six Days in June: How Israel Won the 1967 Arab–Israeli War* (New York: Scribner's, 1992).

Harel, Amos and Avi Isaacharoff, *34 Days: Israel, Hezbollah, and the War in Lebanon* (New York: Palgrave Macmillan, 2008).

Harris, J. P. and F. H. Toase (eds.), *Armoured Warfare* (New York: St. Martin's Press, 1990).

Hecht, Eado and Eitan Shamir, *Medium-intensity Threats: The Case for Beefing-up IDF Ground Forces* (Ramat Gan: The Begin–Sadat Center for Strategic Studies, 2016).

Henkin, Yagil, *The 1956 Suez War and the New World Order in the Middle East: Exodus in Reverse* (Lanham, MD: Lexington Books, 2015).

Herzog, Chaim, *The Arab–Israeli Wars: War and Peace in the Middle East from the War of Independence Through Lebanon*, Rev. ed. (New York: Vintage Books, 1984).

Herzog, Chaim, *The War of Atonement: October, 1973* (Boston: Little, Brown and Co., 1975).

Hoffman, Bruce, *Anonymous Soldiers: The Struggle for Israel, 1917–1947* (New York: Knopf, 2015).

House, Jonathan M., *Combined Arms Warfare in the Twentieth Century* (Lawrence, KS: University Press of Kansas, 2001).

Jones, Clive and Sergio Catignani (eds.), *Israel and Hizbollah: An Asymmetric*

Conflict in Historical and Comparative Perspective (London: Routledge, 2009).

Karsh, Efraim, *Arafat's War: The Man and His Battle for Israeli Conquest* (New York: Grove Press, 2003).

Karsh, Efraim, *The Cautious Bear: Soviet Military Engagement in Middle Eastern Wars in the Post-1967 Era* (Boulder, CO: Westview Press, 1987).

Katz, Samuel M., *Follow Me!: A History of Israel's Military Elite* (London: Arms & Armour Press, 1989).

Katz, Samuel M., *The Ghost Warriors: Inside Israel's Undercover War Against Suicide Terrorism* (New York: Berkley Caliber, 2016).

Kober, Avi, *Israel's Wars of Attrition: Attrition Challenges to Democratic States* (London: Routledge, 2009).

Kreis, John F., *Air Warfare and Air Base Air Defense* (Washington, D.C.: United States Government Printing Office, 1988).

Kurz, Anat and Shlomo Brom (eds.), *The Lessons of Operation Protective Edge* (Tel Aviv: Institute for National Security Studies, 2014).

Lambeth, Benjamin S., *Air Operations in Israel's War Against Hezbollah: Learning from Lebanon and Getting It Right in Gaza* (Santa Monica: Rand Corporation, 2011).

Luttwak, Edward N., *Strategy: The Logic of War and Peace* (Cambridge, MA: Harvard University Press, 1987).

Makovsky, David and Jeffrey White, *Lessons and Implications of the Israel–Hizballah War: A Preliminary Assessment* (Washington, D.C.: The Washington Institute for Near East Policy, 2006).

Matthews, Matt M., *We Were Caught Unprepared: The 2006 Hezbollah–Israel War* (Fort Leavenworth, KS: U.S. Army Combined Arms Center Combat Studies Institute Press, 2008).

McGrath, John J. (ed.), *An Army at War: Change in the Midst of Conflict* (Fort Leavenworth, KS: U.S. Army Command and General Staff College, 2005).

McRaven, William H., *Spec Ops: Case Studies in Special Operations Warfare, Theory and Practice* (Novato, CA: Presidio, 1995).

Mommsen, Klaus, *60 Years Israel Navy* (Norderstedt: Books on Demand, 2011).

Morris, Benny, *1948: A History of the First Arab–Israeli War* (New Haven: Yale University Press, 2008).

Naor, Mordechai (ed.), *Ha'Haganah* (Tel Aviv: Ministry of Defense Publishing House, 1985).

Olsen, John Andreas (ed.), *A History of Air Warfare* (Washington, D.C.: Potomac Books, 2010).

Oren, Michael B., *Six Days of War: June 1967 and the Making of the Modern Middle East* (New York: Oxford University Press, 2002).

Orr, Ori, *These Are My Brothers: A Dramatic Story of Heroism During the Yom Kippur War* (Jerusalem: Contento de Semrik, 2013).

Pollack, Kenneth M., *Arabs at War: Military Effectiveness, 1948–2001* (Lincoln, NE: University of Nebraska Press, 2002).

Pressfield, Steven, *The Lion's Gate: On the Front Lines of the Six Day War* (New York: Penguin, 2014).

Rabinovich, Abraham, *The Boats of Cherbourg: The Navy that Stole Its Own Boats and Revolutionized Naval Warfare* (New York: Henry Holt, 1988).

Rabinovich, Abraham, *The Yom Kippur War: The Epic Encounter that Transformed the Middle East* (New York: Schocken, 2004).

Rabinovich, Itamar and Itai Brun, *Israel Facing a New Middle East: In Search of a National Security Strategy* (Stanford: Hoover Institution Press, 2017).

Rapaport, Amir, *The IDF and the Lessons of the Second Lebanon War* (Tel Aviv: The Begin–Sadat Center for Strategic Studies, 2010).

Rodman, David, *Enforcing the Begin Doctrine: How Israel Stopped Iraq and Syria from Getting the Bomb* (Columbia, SC: CreateSpace, 2017).

Rodman, David, *Sword and Shield of Zion: The Israel Air Force in the Arab–Israeli Conflict, 1948–2012* (Brighton: Sussex Academic Press, 2013).

Rubin, Uzi, *Israel's Air and Missile Defense During the 2014 Gaza War* (Ramat Gan: The Begin–Sadat Center for Strategic Studies, 2015).

Sakal, Emanuel, *Soldier in the Sinai: A General's Account of the Yom Kippur War* (Lexington, KY: University Press of Kentucky, 2014).

Schiff, Ze'ev, *A History of the Israeli Army: 1874 to the Present*, Rev. ed. (New York: Macmillan Publishing Company, 1985).

Searle, Alaric, *Armoured Warfare: A Military, Political and Global History* (London: Bloomsbury, 2017).

Segell, Glen, *Israel, Hamas and Gaza, 14–21 November 2012: Operation Pillar of Defence; Operation Stones of Shale; Operation Blue Sky* (London: The Institute of Security Policy, 2012).

Shalev, Aryeh, *Israel's Intelligence Assessment Before the Yom Kippur War: Disentangling Deception and Distraction* (Brighton: Sussex Academic Press, 2010).

Shalev, Aryeh, *The Intifada: Causes and Effects* (Boulder, CO: Westview Press, 1991).

Shimshoni, Jonathan, *Israel and Conventional Deterrence: Border Warfare from 1953 to 1970* (Ithaca, NY: Cornell University Press, 1988).

Simpkin, Richard E., *Tank Warfare: An Analysis of Soviet and NATO Tank Philosophy* (London: Brassey's, 1979).

Stock, Ernest, *Israel on the Road to Sinai, 1949–1956* (Ithaca, NY: Cornell University Press, 1967).

Turnley, Jessica Glicken, Kobi Michael, and Eyal Ben-Ari (eds.), *Special Operations Forces in the 21st Century: Perspectives from the Social Sciences* (London: Routledge, 2018).

van Creveld, Martin, *The Sword and the Olive: A Critical History of the Israeli Defense Force* (New York: PublicAffairs, 1998).

van Creveld, Martin, Steven L. Canby, and Kenneth S. Brower, *Air Power and Maneuver Warfare* (Maxwell Air Force Base, AL: Air University Press, 1994).

Williams, Louis (ed.), *Military Aspects of the Arab–Israeli Conflict* (Tel Aviv: University Publishing Projects, 1975).

Yaniv, Avner, *Deterrence Without the Bomb: The Politics of Israeli Strategy* (Lexington, MA: Lexington Books, 1987).

Yonay, Ehud, *No Margin for Error: The Making of the Israeli Air Force* (New York: Pantheon, 1993).

Documents

Central Intelligence Agency. Directorate of Intelligence, *The 1973 Arab–Israeli War: Overview and Analysis of the Conflict* (Langley, VA: Central Intelligence Agency, 1975).

Government of Canada, *Canadian Special Operation Forces Command* (Ottawa: Department of National Defense, 2008).

Joint Chiefs of Staff, *Joint Publication 3–05 (Test): Doctrine for Joint Special Operations* (Washington, D.C.: Office of the Joint Chiefs of Staff, 1990).

Joint Technical Coordinating Group for Munitions Effectiveness, *Special Report: Survey of Combat Damage to Tanks*, Vol. I–III (Washington, D.C.: Defense Intelligence Agency, 1970).

Swain, Richard M., Donald L. Gilmore, and Carolyn D. Conway, *Selected Papers of General William E. DePuy: First Commander, U.S. Army Training and Doctrine Command, 1 July 1973* (Fort Leavenworth, KS: U.S. Army Command and General Staff College, 1994).

The Database Desk, Incidents and Activists Database, *Operation 'Protective Edge': A Detailed Summary of Events* (Herzliya: International Institute for Counter-terrorism, 2014).

United States Army. Department of Army Headquarters, *Army Doctrine Reference Publication No. 3–0: Unified Land Operations* (Washington, D.C.: Department of the Army Headquarters, 2012).

Various documents from the United States National Archives, College Park, MD.

Personal Communications

Israel Air Force (IAF) Brigadier General (Ret.) Itzhak Amitay, June 25, 2007.

Israel Air Force (IAF) Lieutenant Colonel (Ret.) Amos Cohen, August 10, 2008.

Web Sites

From the Grapevine: <www.fromthegrapevine.com>.

Israel Air Force (IAF): <www.iaf.org.il>.

Israel Between the Lines: <israelbetweenthelines.com>.

Israeli Knesset: <www.knesset.gov.il>.

Yediot Ahronot: <www.ynetnews.com>.

Index

About the Author

David Rodman is the author of five previous books about Israeli diplomatic and military history. He has also published articles in academic journals, including *Intelligence and National Security*, *The Journal of Military History*, *The Journal of Strategic Studies*, *Middle Eastern Studies*, *Israel Affairs*, *MERIA Journal*, *Defence Studies*, and *The Israel Journal of Foreign Affairs*.

Printed and bound by CPI Group (UK) Ltd, Croydon, CR0 4YY

13/09/2026

14961102-0003